On Our Minds

PUBLISHING FOR THE WORLD
125 Years

THE JOHNS HOPKINS UNIVERSITY PRESS

On Our Minds

How
Evolutionary Psychology
Is Reshaping the
Nature-versus-Nurture
Debate

ERIC M. GANDER

The Johns Hopkins University Press
Baltimore & London

© 2003 The Johns Hopkins University Press
All rights reserved. Published 2003
Printed in the United States of America on acid-free paper

2   4   6   8   9   7   5   3   1

The Johns Hopkins University Press
2715 North Charles Street
Baltimore, Maryland 21218–4363
www.press.jhu.edu

Library of Congress Cataloging-in-Publication Data
Gander, Eric.
On our minds: how evolutionary psychology is reshaping the
nature-versus-nurture debate / Eric M. Gander.
p.   cm.
Includes bibliographical references and index.
ISBN 0–8018–7387–8 (hardcover : alk. paper)
1. Genetic psychology. I. Title.
BF701 .G26 2003
155.7—dc21
2002152158

*A catalog record for this book is available
from the British Library.*

The desire to dedicate the fruits of our labors to those we most love and respect is surely a basic aspect of human nature.

This book is dedicated to my wife,

Lauralee

Descended from the apes! My dear, let us hope that it is not true, but if it is, let us pray that it will not become generally known.

—Attributed to the wife of the Bishop of Worcester, after the publication of *The Origin of Species* (1859)

"The greatest magician (Novalis has memorably written) would be the one who would cast over himself a spell so complete that he would take his own phantasmagorias as autonomous appearances. Would not this be our case?" I conjecture that this is so.

—Jorge Luis Borges, "Avatars of the Tortoise," 1952

CONTENTS

On Our Minds

INTRODUCTION

"This Changes Everything"

Since turning thirty, I have been a committed "conservative" in at least one respect. I hold fast to the maxim that, as the French say, *plus ça change, plus c'est la même chose.* This is not always a popular, or even a very American, idea. On the contrary, there are many who will tell you that one or another invention, idea, or world-historical event changes everything. Perhaps it is the invention of the internet that changes everything by changing that all-important human activity of communication. Or perhaps it is quantum mechanics that changes everything by changing that all-important human activity of understanding our physical world. Or perhaps it is the Holocaust that changes everything by changing that all-important human activity of coming to grips with good and evil. Or perhaps it is a moon landing, a brutal terrorist attack, or an especially deadly virus that changes everything. Perhaps. But I suspect that those who say "this changes everything" may be guilty of special pleading on the part of their particular interest. Intellectuals, academics, and authors may be especially guilty of this type of special pleading. Indeed, my bookshelves groan under the weight of books by authors who boldly proclaim of their subject matter: "This changes everything." But is the change heralded even by the best of these books really fundamental? Or is it merely a surface kind of change?

Imagine a Roman living some nineteen hundred years ago, around the time Hadrian completed the Pantheon. Rome was then an empire of considerable military and political breadth. It could boast of engineering marvels like the aqueducts, and its provinces were home to many different philosophies and belief systems—one of which based its central tenets on the teachings of a crucified Jewish radical. Certainly no one could deny that the ancient Roman empire changed the world in an uncountable infinity of ways. Thus our historical Roman would clearly be justified in looking to numerous aspects of his homeland and saying: "This changes everything." But even as the world changed, it stayed much the same.

Imagine that our Roman is also a gifted seer, who can peer ahead into the twenty-first century. He would see change on a massive and unimaginable scale. But he would *also* see a century riven by wars, fought for the same reasons that wars have always been fought: territorial conquest, the control of resources, and the establishment of new hierarchies. He would see a world in which the political and military spheres of life are still dominated by high-status males, almost all of whom are susceptible to the same vices (and virtues) as the Roman senators or Caesars of old. Indeed, the twenty-first century would furnish him with quite a long list of alpha-male politicians who could not seem to keep their hands off of young, fertile females. At the same time, of course, his list would also include young, fertile females who found themselves irresistibly attracted to older, high-status males whom they would attempt to seduce whenever possible. More generally, our Roman seer would find in the twenty-first century a world in which hierarchy is ubiquitous, where control of resources matters and is often the source of conflict, where women and men continue to want different things from life, and where nepotism, sibling rivalry, and adolescent rebellion—but also altruism, bravery, and the love of parents for their children—are all commonplace. Cars may have replaced chariots, but I suspect that ancient Roman teenage boys were likely driving their chariots too fast and thereby risking unnecessary injury, all in a desperate attempt to impress some young lady. In sum, although a great deal has changed, a great deal has remained the same.

So the next time someone says, "This changes everything," think for a moment about the one thing that never changes even when everything else does: human nature. Now, I realize that this claim about the persistence—let alone the existence—of human nature is quite controversial in some circles. I also realize the obvious relatedness of this claim to the debate concerning the roles of biology and culture in explaining human affairs. This is the perennial nature-versus-nurture debate—a debate that was old when the ancient Greeks conceptualized it as a conflict between *phusis* (nature) and *nomos* (law, custom, or convention).

With respect to this debate, it is safe to say that for much of the past century human nature endured a fairly sustained period of exile. During this time the prevailing belief seemed to be that humans are the way we are *solely* because of the cultural influences on us. In short, our upbringing or education accounted for everything in determining what would become of any one of us. This view is still embraced by many intellectuals, including the contemporary philosopher and social critic Richard Rorty, who insists that "socialization, and thus historical circumstance, goes all the way down . . . there is nothing 'beneath' socialization or prior to history which is definatory of the human."[1]

But despite their influence in humanities departments of American universities, Rorty and his intellectual colleagues seem strangely out of touch with the mind-set of contemporary culture. Indeed, if you have even a passing familiarity with the recent torrent of articles and best-selling books written by scientists for an audience of educated nonscientists, you cannot help noticing it: Human nature is back. The thinking now seems to be that a complex and richly detailed human nature really does exist, that it is to a very large degree scientifically knowable, that it differs markedly between the sexes, that it delimits a set of viable human cultures, and that, because of all this, it makes a big difference when we set out to discuss moral, ethical, and political questions.

The return of human nature has been facilitated by the emergence of a branch of science now known as evolutionary psychology. Succinctly put, evolutionary psychology is an interdisciplinary science that attempts to understand how the human mind works by viewing the mind as—in the words of Steven Pinker, a leading evolutionary psychologist—"a system of organs of computation, designed by natural selection to solve the kinds of problems our ancestors faced in their foraging way of life, in particular, understanding and outmaneuvering objects, animals, plants, and other people."[2] These organs of computation are sometimes called *mental modules* by evolutionary psychologists. Apparently we have mental modules that enable us to perform an enormously wide variety of tasks, including selecting a mate, deciding what amount of our resources to invest in our various children, understanding how the minds of other individuals work, recognizing faces, rotating images in our minds, and detecting when someone is trying to cheat us.

To the extent that culture is created by collections of evolved individual minds working in some degree of unison, evolutionary psychologists claim special insight not only into how cultures are generated but also into which cultures are humanly possible. The phrase *evolutionary psychology* itself came into widespread use as the result of an enormously influential volume of essays entitled *The Adapted Mind: Evolutionary Psychology and the Generation of Culture,* edited by Jerome H. Barkow, Leda Cosmides, and John Tooby. As the editors of that volume explain, "Evolutionary psychology is simply psychology that is informed by the additional knowledge that evolutionary biology has to offer, in the expectation that understanding the process that designed the human mind will advance the discovery of its architecture. It unites modern evolutionary biology with the cognitive revolution in a way that has the potential to draw together all of the disparate branches of psychology into a single organized system of knowledge."[3]

The critical point here is that evolutionary psychology understands the human mind not as an essentially blank slate upon which culture writes its dictates, nor as a mysterious vessel that now contains the essence of our humanity (an essence that may once have been thought to reside in the soul).[4] Rather, evolutionary psychology understands the mind as simply another part of the human body, albeit an especially complex part. Like all parts of the body, the mind has a specific function: information processing and computation. Also, according to evolutionary psychologists, the human mind, like the body, must have evolved over the course of the last two million or so years of humanoid evolution.

This understanding of the mind is simultaneously appealing and distressing. It is appealing because it argues for the overall psychic unity of mankind and womankind. It suggests that underneath the outwardly different and sometimes bizarre cultures that anthropologists tell us exist and have existed on the planet earth, men and women are now, and have been for at least the past one hundred thousand years, pretty much the same everywhere. Members of each sex display the same basic patterns of emotional reactions and reasoning processes, the same desires for the same types of physical and social rewards, and the same attitudes toward others and toward the physical world. The hundred thousand years figure, by the way, comes from the fact that, given the glacially slow pace of humanoid evolution, the human mind itself has not changed appreciably from what it was structurally one hundred thousand years ago.

But this understanding is also distressing because it strongly suggests that the human mind as it exists today may be tragically ill-equipped to deal with the problems faced by modern humans. After all, our hunter-gatherer ancestors of one million, or even one hundred thousand, years ago never faced the problems attendant to noisy, overcrowded urban population centers. Additionally, they never needed to compute probabilities concerning situations that occurred beyond the realm of their small foraging group.

Thus it is no wonder that the modern mind seems confused. Consider the September 11 terrorist attacks. Terrorists know that the salience the mass media produce can cause individuals to miscalculate the probability that a particular fate will befall them. One hundred thousand years ago, if you saw fire and devastation, you could be sure you were probably in the thick of it, and you needed to react accordingly. Today, you can view a catastrophe on your television set, a half a world away from where it may be happening. That is precisely why the point of terrorism is to garner as much attention as possible—something that is not difficult to accomplish given the reach of the modern media in our global village. This attention gives terrorists a type of power vastly out of proportion to their actual numbers. The problem here is not with the mass media. The problem is that the information-processing and computational device that is the evolved human mind, although very complex, is still attuned to an environment much different from the one in which most individuals today find themselves.

Ironically, the widespread appeal of evolutionary psychology—evident even in the academy and growing with the publication of each new book on the subject—may be due in part to its claims to be able to explain some of the more problematic aspects of the human condition. Any theory that appears even mildly successful at explaining why suicidal vengeance persists in our world, or why some women sometimes abandon their newborn infants, or why equality between the sexes seems so difficult to achieve, will find a ready audience today.

But there is another, more important, reason for evolutionary psychology's widespread and growing appeal. We can get a clue to this reason by reflecting on all this talk of the *evolved* mind as an *information-processing and computational device* that can now be understood in a *scientific* manner. We notice here that the language of evolutionary psychology is strikingly au courant. It is the language of science, computers, and genes. This is the language of the twenty-first century. This helps to explain why evolutionary psychologists are everywhere, offering an explanation for anything and everything. Just some of the questions addressed by evolutionary psychologists include: why we love flowers (apparently we needed to attend to flowering plants in the ancestral environment); why we get seasick (the answer may surprise you); why men buy women expensive diamond engagement rings from overpriced retail jewelers; why there are more men than women CEOs; why parents sometimes invest more in their female children than in their male children; why we get so upset when the person ahead of us in the express checkout line of the super-market has one or two more than the allowed number of items in his or her shopping cart; why we are kind to strangers; why our social, cultural, and political institutions work the way they do; and, in general, why humans are the way we are.

As with any good science, evolutionary psychology's explanatory reach certainly appears to exceed its grasp—for the time being, at least. But for just this reason we may wonder whether some of the questions posed in the previous paragraph are appropriate for evolutionary psychology or, indeed, for *any* science. Evolutionary psychologists answer, resoundingly, that the examination of these questions is appropriate for their discipline. They also assert that only real science can provide satisfactory answers to such ques-tions. For evolutionary psychologists, real science must conform to the model of what is done in natural science departments in universities—departments like mathematics, physics, chemistry, biology, and cognitive science.

In making this argument evolutionary psychologists are quick to note that during much of the past century—indeed for much of the time that human nature was in exile—both the humanities (especially philosophy) and the so-called social sciences (including anthropology, sociology, and psychology it-self) have made a muddle of their investigations of the human condition, resulting in politically correct endorsements of sixties-style social engineering as well as the various doctrines of "cultural relativism" that are today the staple of education in the humanities and social sciences. Such endorsements and the doctrines that ground them—doctrines that tell us, for example, that socialization goes all the way down and that humans can therefore be indoc-trinated or "educated" to thrive in almost any imaginable culture—are quite appealing to individuals who are better at manipulating words than genes. These doctrines "empower" those of us in the humanities and social sciences to think we can solve all human problems by doing what we love most: working with and studying language. Indeed, if languages are the foundations of cultures, and cultures exclusively shape who we are, then changing our language—that is, changing the very words and metaphors we use—must change us in the only *fundamental* sense possible.

Beyond its appeal to many in the humanities and social sciences, this view of the human condition—a view that privileges the nurture side of the nature-versus-nurture debate—finds a receptive audience among many pragmatists and idealists. Pragmatists find it appealing because, for the time being at least, "language engineering"—of the type that is accomplished by the right "education," or by public service announcements reminding us to be careful of what we say around our children, or, failing all this, by speech codes is easier than genetic engineering. (Such language engineering is grounded in the belief that we can, for example, eliminate racist thoughts by eliminating the ability to use racist language.) Idealists find this view appealing because it implies that the only limit to human cultures is the limit of our imaginations, as those imaginations are used to form new words and metaphors. In this view, utopian vistas stretch before us, infinite and boundless, generated courtesy of the poets, who really are the legislators of the world.

It is a dour soul who would gainsay this view. Certainly evolutionary psychologists like Pinker and the editors of *The Adapted Mind* have no desire to appear less optimistic about the human condition than their colleagues in the humanities and social sciences. But many evolutionary psychologists (especially Pinker) are adept enough at persuasion to see they face a dilemma. By seeming to argue for a materialistic human nature that can be known by scientific methods, they seem to be arguing for *limits* on human cultures, thus denying the possibility that we can re-create ourselves in an infinite number of ways. This makes them appear duller and less optimistic about the human condition than the poets. On the other hand, evolutionary psychologists cannot simply abandon science and cast their lot with the poets, for this would make them appear nonrigorous, thus demolishing their claim to be a real science.

In the face of this dilemma, evolutionary psychologists insist that the whole nature-versus-nurture debate must be transcended. This transcendence comes by seeing that although human cultures are in some sense bounded by human nature, the set of viable human cultures is still infinite. In other words, the set of viable human cultures is best described as a bounded infinite set.[5] Mathematicians are quite familiar with such sets, but the example I shall use is not drawn from formal mathematics. Consider the set of all grammatically correct English sentences. Notice that this set is infinite.[6] Now consider the set of all strings of English words, where the first word in the string is capitalized and the last word in the string is followed by a period. Notice that this set is also infinite. Now notice that every member of the first set is in the second set, but not every member of the second set is in the first set. Thus the set of grammatically correct English sentences is infinite but bounded by the rules of English grammar. Also, it is precisely the boundaries on this set that allow any of its members to convey any meaning.

As it is with grammatically correct English sentences, so it is with cultures. Thus you may find a culture in which husbands are allowed only one wife, and you may find another culture in which husbands are allowed three wives. You may even find a culture in which husbands are allowed hundreds of wives. Similarly, you may find some cultures in which wives have no legal rights apart from their husbands, and you may find other cultures in which wives and husbands have equal legal rights. But you will never find a *viable* human culture without husbands and wives.[7]

The idea that human cultures form a bounded infinite set may help to explain the enduring and universal appeal of great works of literature. However remote in time and space such works may be from us, the themes and characters contained therein are nonetheless accessible because we all share a common human nature. We can *identify* with what is going on in such works. This is surely part of what W. E. B. Du Bois was emphasizing when he wrote: "I sit with Shakespeare and he winces not. Across the color line I move arm in arm with Balzac and Dumas, where smiling men and welcoming women glide in gilded halls. From out the caves of evening that swing between the strong-limbed earth and the tracery of the stars, I summon Aristotle and Aurelius and what soul I will, and they come all graciously with no scorn nor condescension."[8]

But before we get too caught up in this unbridled celebration of the bounded infinite set of viable human cultures open to us all, we should remind ourselves of the point of contention between the evolutionary psychologists and the poets. The view of human nature and human culture offered by the evolutionary psychologists still means that there are some *imaginable* cultures—for example, cultures *without* status hierarchies, or resource competition, or material "possessions"—that simply are *not* viable. The return of human nature means that the debate about the boundaries human nature places on human cultures is once again joined. Evolutionary psychologists say they have something valuable to contribute to that debate. I want to examine critically their contribution.

This is a book about the contemporary public discourse surrounding human nature and evolutionary psychology. It is written by a student of public discourse and public argument. I define *public discourse* broadly, as debate and discussion about significant ideas, issues, values, policies, and the like, directed for the most part at an educated but nonspecialized audience. Such discourse may emanate from various arenas, including the press and the mass media, the academy, and the institutions of government. But its audience must be "the public." My principal aim in what follows is to sharpen public discourse about human nature and evolutionary psychology. I intend to do this largely by examining the popular-science writings of evolutionary psychologists and other natural scientists, as well as related works by social scientists, philosophers, rhetorical and cultural critics, ethicists, and others. All of the principal works I examine are consciously written for the general public. This book is thus a synthesis of ideas. I attempt to survey a large intellectual terrain with the intention of understanding how certain ideas related to science, evolutionary psychology, and human nature are discussed and debated in contemporary culture.

The approach that I take toward the works that I will be examining—particularly the works of popular science—is one that I would describe as constructively critical. This approach seeks to understand in a *commonsensical* way what the authors of these works are trying to say. This is not always the approach that many in the humanities and social sciences today take toward scientific writings.[9] The academic scandal caused by Alan Sokal illustrates this point. In 1996, the editors of *Social Text,* one of the premier journals in the field of cultural studies, published a special double issue on science. Contributors included a number of luminaries in the field of science studies, and a lone physicist named Alan Sokal, who wrote an article entitled "Transgressing the Boundaries: Toward a Transformative Hermeneutics of Quantum Gravity."[10] Sokal's article, written in a style one might describe as high postmodernism, advanced the argument that "postmodern science" now validates postmodern literary theory, and all the world is really a text. For example, Sokal offers an analysis of an observation by Jacques Derrida that the "Einsteinian constant is not a constant" but, rather, "the very concept of the game."[11] Sokal then writes:

> In mathematical terms, Derrida's observation relates to the invariance of the Einstein field equation $G\mu\upsilon = 8\pi GT\mu\upsilon$ under nonlinear space-time diffeomorphisms (self-mappings of the space-time manifold that are infinitely differentiable but not necessarily analytic). The key point is that this invariance group "acts transitively": this means that any space-time point, if it exists at all, can be transformed into any other. In this way the infinite-

dimensional invariance group erodes the distinction between observer and observed; the π of Euclid and the G of Newton, formerly thought to be constant and universal, are now perceived in their ineluctable historicity; and the putative observer becomes fatally de-centered, disconnected from any epistemic link to a space-time point that can no longer be defined by geometry alone.[12]

Sokal's article was provocative and transgressive. But there was one problem with it—a problem that Sokal revealed immediately after its publication.[13] The article was a hoax. To be sure, none of the quotations used by Sokal were fabricated. These quotations—from works by Derrida, Robert Markley, Stanley Aronowitz, and others—were all rigorously accurate, meticulously documented, and entirely representative of the works from which they were taken. But the accuracy with which these quotations were rendered was the only rigorous part of the article. The rest was an imaginative mix of mathematical gibberish, aggressive nonsense, postmodern jargon, and moralistic calls for a libratory science and politics that undermines capitalism, patriarchy, militarism, and so forth. Sokal bet that he could write an article parodying the language of science studies and have that article accepted for publication by a leading journal in the field of cultural studies without the editors of the journal recognizing the article as a parody or even raising an eyebrow over his astonishing claim that π is not a mathematical constant.

Sokal's transgressive act sent shock waves through the academic world. But it teaches at least two valuable lessons. One is that the natural sciences do not need to be dressed up in fancy postmodern garb; they are interesting enough in a plain white lab coat. The truth is that noted physicists and popular-science writers like Steven Weinberg and Stephen Hawking have more interesting things to say about Einstein than Derrida ever will—a comment that in no way disparages Derrida's brilliant readings of literary and philosophical figures like Rousseau and Plato. And what I have said of Weinberg and Hawking on physics could well be said of Pinker on evolutionary psychology. It is virtually impossible *not* to be fascinated by discussions of how the mind creates language by using words and rules, or how the mind equips us to detect even very subtle instances of when we are being cheated, but not to detect instances of when we are being aided.

The second lesson that Sokal's hoax teaches is that "postmodern" discourse tends to obscure rather than clarify *public* discourse about science. Ironically, while postmodern discourse is frequently spoken in the name of, or for the benefit of, the public, it often seems consciously constructed to make it as incomprehensible as possible to the public. One might conclude that many of those in cultural studies and science studies have all but given up on the public. But such pessimism regarding the public's ability to understand vital issues—particularly with respect to science—does no one any good. Nor should it be encouraged, especially today when scientific discoveries and the theories that relate to them—including those offered by evolutionary psychologists—are becoming increasingly important to our individual and collective lives. Thus I would insist that engaging and analyzing the popular-science writings of evolutionary psychologists directly and commonsensically can be a productive exercise in reaffirming the value of public discourse in a democratic society.

To do this we must understand what evolutionary psychologists mean when they use terms like *mental module, adaptation, selection pressure,* and so forth. Once this is done we can then proceed to analyze and critique attempts by evolutionary psychologists to "reverse engineer" the mind—that is, attempts to explain our mental modules as adaptations to specific selection pressures. We can ask whether we can ever have enough information about the ancestral environment to make accounts of the evolution of mental modules anything more than "just so stories." More specifically, we can ask how the ability to write poetry in iambic pentameter or to solve a first order differential equation could have evolved, since there seems no indication of how either talent could have conferred a survival or reproductive advantage on any of our hunter-gatherer ancestors. Finally, we can inquire about the ways in which evolutionary psychology must inevitably change our understanding of, and discourse about, ourselves and our world. We can—indeed we must—ask whether our contemporary, results-oriented understanding of equality between the sexes can be made consistent with what evolutionary psychology is telling us about the significant emotional and intellectual differences between men and women. Similarly, we will need very soon to examine closely our understanding of moral and ethical responsibility in light of what evolutionary psychology is telling us about the "hardwired" nature of our emotions and desires. Evolutionary psychology even invites, and perhaps enables, us to ask anew the oldest questions of political philosophy. What is the relationship between the individual and the community? Is there an understanding of justice natural to humans that could compel us to act for a good beyond our own immediate self-interest or the interest of our kin? Or does "survival of the fittest" really condemn us to a world devoid of altruism of any kind?

I am not going to pretend to provide answers to these or other related questions. Rather, I want simply to frame these questions intelligently and to engage them seriously. Fortunately, this task is made manageable and even enjoyable by the fact that many contemporary natural scientists whose work I shall discuss are gifted science writers and capable advocates for their various positions. This may well explain why many are also best-selling authors. Indeed, evolutionary psychologists like Steven Pinker and Geoffrey Miller, together with natural scientists from related disciplines—individuals like Richard Dawkins, Sarah Blaffer Hrdy, and Edward O. Wilson—have all produced important popular-science books dealing directly with the concerns of evolutionary psychology.

I should hasten to add that in addition to these and many other well-known advocates of evolutionary psychology, those who are not in full agreement with this new science also number among their group some of the best science writers of our time. Richard Lewontin, Steven Rose, and the late Stephen Jay Gould would all be good examples of leading scientists and engaging writers who effectively present forceful critiques of various claims made by evolutionary psychologists. Indeed, among natural scientists, Stephen Jay Gould may well have been the most gifted popular-science writer since the likes of Isaac Asimov and George Gamow.

Thus we have with respect to evolutionary psychology a spirited debate about a matter of vital public concern (the nature of human nature), carried on largely by scientists who direct much of their discourse to a general audience of educated nonspecialists who could well be said to comprise the illusive public. If we remember that Charles Darwin, like almost all of his nineteenth-century colleagues in science, directed his most important works to the lay readers of his time, we might feel justifiably proud that our era has been able to maintain the tradition of popular-science writing.

My task of sharpening public discourse about human nature and evolutionary psychology is also made urgent by the fact that the natural sciences have finally caught up with more than two thousand years of "mere" philosophical speculation about human nature, and are now poised to surpass it. We have already decoded the human genome. We are daily finding new connections between genes and the evolved minds that they build. During the mid-1990s, significant media attention was focused on the discovery of the so-called gay gene that may predispose individuals toward homosexuality. Aside from behavioral traits like sexual preference, genes have also been linked to specific mental impairments like autism and attention deficit hyperactivity disorder (ADHD). More generally, there is significant evidence, coming from studies on monozygotic (i.e., genetically identical) twins who have been reared apart, that traits like temperament, religiosity, and general intelligence are all highly heritable and (for reasons that I shall be at pains to explain at length) largely genetically based.

Of course, to speak of a genetic basis for traits like intelligence is immediately to raise issues of group differences in IQ test scores and nonoverlapping bell curves. Evolutionary psychologists, particularly the younger ones like Pinker, go to considerable lengths to deny that their work has any practical or theoretical connection to the questions raised by the research into these issues. But such denials seem ultimately futile. Still, while some of this research may at first glance seem to aid in advancing a conservative—or even a reactionary—political agenda, the situation is not that simple. In fact, cultural conservatives and religious fundamentalists who believe it to be God's will that, for example, women remain at home while their husbands earn the families' daily bread, and who attempt to use the work of evolutionary psychologists to support this belief, may rue the day they first heard of mental modules. The devout scientific materialism of evolutionary psychology is in fairly strong tension with the mystical elements of religion, or ultimately with the belief that there is a will or a God. Hence those who attempt to use evolutionary psychology to advance any specific political agenda may find that they have, in a manner of speaking, grasped a genetically engineered tiger by the tail.

In a sense, the real "danger" posed by an embrace of evolutionary psychology cuts across almost all political and philosophical systems. It is the danger manifested in a reductionistic view of the human self. It is a danger that is also the promise and the essence of science. Once we understand the mind as an organ evolved for information processing and computation, we will be much better able to predict and control the mind. But this may come at the expense of not being able to see ourselves as anything "higher" or more majestic than an arbitrary arrangement of amino acids and the proteins that they code for.

Indeed, the field of applied cognitive science has already made fascinating advances in "reducing" one important aspect of the self—the very concept of intelligence—to measurable brainwave activity. In the early 1990s, scientists at a U.S. firm called Neurometrics, using research and data published in the 1980s by neuroscientists, developed a device for measuring human intelligence that they dubbed the "IQ Cap." The device consisted of sixteen electrodes, a standard electroencephalograph (used for measuring brainwave activity), and a computer. The electrodes were attached to a test subject's head in a manner that barely disturbed the subject's coiffure. Then the subject was asked to stare at a distant object, such as a doorknob or a thumbtack stuck to a wall. In less than thirty seconds the subject's brainwaves were analyzed and the computer registered a number that predicted, within one-half of a standard deviation, what the subject would score on the Wechsler Adult Intelligence Scale—the most comprehensive IQ test available today. The IQ Cap even worked as well on predicting what children would score on the Wechsler Intelligence Scale for Children.[14] Oddly, Neurometrics was unable to market the IQ Cap successfully, and later went bankrupt. Tom Wolfe, the novelist and journalist, persuasively argues that the failure of Neurometrics had nothing to do with technology and everything to do with our contemporary cultural mind-set. In an article entitled "Sorry, Your Soul Just Died," he writes: "It wasn't simply that no one *believed* you could derive IQ scores from brainwaves—it was that nobody *wanted* to believe it could be done."[15] I think that both the popularity of—and the controversy surrounding—popular-science books on evolutionary psychology suggests that it is still an open question as to how many of us today *want* to believe that the mind can be explained as an evolved organ of information processing and computation. This means that there may well be rough waters ahead as we try to come to grips with what the evolutionary psychologists are attempting, through their popular-science writings, to tell us about how our minds work and who we are.

To help navigate these waters, this book is divided into three parts. Part one is historical and contextual. It attempts to "place" evolutionary psychology within the context of biological explanations of the human mind and of human cultures, particularly as these explanations have emerged since the nineteenth century. In this part, I also pay special attention to the development and critical reception of sociobiology, the most immediate predecessor to evolutionary psychology. Part two provides an in-depth discussion of the central principles of evolutionary psychology. Finally, part three examines the connection between human nature and human culture, in light of the claims and theories advanced by evolutionary psychologists.

By now I suppose that the discerning reader will have guessed where this introduction must end. If what I have said thus far is true—indeed if evolutionary psychology makes good on only a fraction of the promises its promoters have advanced—then this really does change everything, and change it in a way that is truly fundamental. The reason is simple. Self-knowledge is the only knowledge that could ever lead to true change. All other knowledge is merely incidental. Thus the new science of the mind that has crystallized around evolutionary psychology may represent the best attempt humankind has yet undertaken to fulfill the Delphic injunction "know thyself." At the very least, evolutionary psychology will continue to provide fascinating insights into the structure and function of various mental modules. More ambitiously, evolutionary psychology will help us to know better the relationship between human nature and human culture, so that the latter can be made to accord better with the former.

But there's the rub. Why must culture be changed to accord with human nature, rather than human nature being changed to accord with the various cultures that we want to exist on this planet? The answer is that we cannot yet alter our fundamental human nature. That *will* eventually change. We can glimpse the beginnings of that change already. At its most ambitious, then, evolutionary psychology will be a major player in helping to bring about, through scientific means, alterations in our fundamental human nature and, thus, fundamental alterations in future human cultures and human history. Such cultural and historical alterations will be of a scope and magnitude that sixties-style social engineering, or even contemporary attempts at mass persuasion or propaganda, could never come close to achieving, at even their most effective levels. The final irony is that when this change happens we may no longer *be* human, and all of what you are about to read—indeed all of evolutionary psychology up to that point and all of science, art, religion, philosophy, and literature—may be to our "descendants" simply the prehistory of their new species. But we have not reached that point yet. Hence, for the time being, it is important that we try to understand our human nature, while we still have it.

PART ONE

The Evolution of a Controversy

Inheritance plays an important part in determining mental performance. It is my own conviction that the arguments of the environmentalists are too much based on sentimentalism. They are often even fanatic on this subject. If the facts support the genetic interpretation, then the accusation of being undemocratic must not be hurled at the biologists. If anyone is undemocratic on this issue, it must be Mother Nature.

—L. L. Thurstone, "Theories of Intelligence," 1946

CHAPTER ONE

Stephen Jay Gould Historicizes Science

Evolutionary psychology is a *real* science. At least, that is what its proponents claim. But evolutionary psychology just is *not* theoretical physics or organic chemistry. It may be true to say that its methods are those of the natural sciences, but it is difficult to say that evolutionary psychology's object of study is *merely* the natural world. Indeed, as Steven Pinker points out, evolutionary psychology's object of study cannot even properly be said to be the human *brain*.[1] Rather, its object of study is properly said to be nothing less than the human *mind* in all its richness and complexity. But just how does the human mind study *itself*? Therein lies the problem.

Given its focus on the human mind, it would not be unfair to describe evolutionary psychology as the latest attempt to develop a modern "science of man." Such a science must of necessity treat moral and ethical matters, for such matters are what distinguish humans as humans from other animals. But, as critics of evolutionary psychology are quick to point out, the idea that modern science can inform us about our deepest moral and ethical questions by examining the physical self (which, after all, is all modern science has to work with), though as old as modern science itself, has never met with much success and indeed has often tended to say more about the scientist doing the examining than about anything he or she was investigating. These same critics are equally quick to point out that a historical knowledge of science tends to undermine our faith in scientific pronouncements about our humanity, particularly those that focus on the moral or ethical capacities of humans, or those that purport to demonstrate moral, ethical or, especially, intellectual differences between classes of humans. As the critics say, if history teaches us anything it is that even the best and most respected scientists have never been wholly immune to the political and ideological forces of their day. This well accounts for the abundance of past "scientific" theories that "proved" the inferiority of one race to another or that "proved" the hysterical nature of women. The fact that such theories have all been discredited today is reassuring, the critics will admit. But it is also evidence of the need for continued vigilance against the encroachment of science into domains where it does not belong. The separate but related cases of two prominent nineteenth-century scientists, Samuel Morton and Paul Broca, may serve as extended examples of just this point.

Morton, a physician by training and a research scientist by vocation, lived in Philadelphia during the early and mid–nineteenth century. He was so well respected as a scientist that upon his death in 1851 the *New York Tribune* wrote that "probably no scientific man in America enjoyed a higher reputation among scholars throughout the world, than Dr. Morton."[2] It would be ghoulish, but not unfair, to say that Morton's reputation as a scientist rested on a foundation of over a thousand human skulls. It would also not be unfair to say that what he did with those skulls may have made him one of the best friends a Southern slaveholder could have.

As part of the research for his enormously popular books (including *Crania Americana*, 1839, and *Crania Aegyptiaca*, 1844), Morton took measurements of several hundred skulls that had belonged to individuals of different races. He "scientifically" established that, on average, whites had the biggest skulls, American Indians the next biggest, and blacks the smallest. Indeed, Morton's work was so meticulous he even established that, among whites, Anglo-Saxons had bigger skulls than Jews or "Hindoos," i.e., individuals from the Indian subcontinent. Of course no one was at all surprised by Morton's discoveries, which were taken as scientific proof of the intellectual superiority of whites over those of other races, and which were quickly used to justify the political and social practices of the day. The assumption was that those who had big skulls had big brains, and those who had big brains were more intelligent than those with smaller brains. From this, everything else followed. Indeed, Morton's work led one of the South's most respected medical journals to print this acknowledgment: "We of the South should consider [Morton] as our benefactor, for aiding most materially in giving to the negro his true position as an inferior race."[3]

Today it is all too easy to discredit any current scientific theory that has troubling implications for morality or ethics by *historicizing* it. One simply notes that past theories, like those of Morton and his colleagues, have often amounted to nothing more than elaborate justifications for oppression. Then one confidently proceeds to draw the historical conclusion that the present theories we do not like will doubtless suffer in the future the same fate as those that once lent support to past oppressors. Indeed, since the bad theories of the past (like those of Morton) were once popular, and since bad theories of today that we seek to discredit must also be popular (who needs to discredit an unpopular theory?), one cannot help but feel a special kind of superiority when discrediting bad popular theories. Doing so puts one on the right side of history after all, even if that fact will not be revealed until well after everyone who currently supports or opposes the theory in question is dead.

But this very popular way of dealing with scientific theories that we do not like is simply not sufficient. Indeed, it makes a difference how we discredit or invalidate a scientific theory, for exactly the same reason that it makes a difference how we validate a scientific theory. In fact, although valuable in its own right, cultural, historical, or political knowledge is not particularly helpful in evaluating the truth of scientific theories as such. Indeed, I would argue that a proper historical understanding of science reveals that science itself provides the best corrective to its own mistakes.

To help support this claim, and to get back to Doctor Morton and his skulls, I want now to turn to an examination of the central argument in Stephen Jay Gould's 1981 book *The Mismeasure of Man*. When he died in 2002, Gould held many titles and positions, including Alexander Agassiz Professor of zoology and professor of geology at Harvard University. But I strongly suspect that at the time of his death Gould took the most professional pride in being probably the best, and certainly the most beloved, popular-science writer of his time. The numerous books and articles Gould wrote for public consumption brought the insights of evolutionary theory to bear on a very wide range of scientific and nonscientific issues. *The Mismeasure of Man* is, in fact, a perfect example of the type of book that made Gould simultaneously popular among a lay audience, respected by his scientific colleagues, and more than a bit controversial. The book presents science in an interesting and understandable way, while at the same time forcefully advocating a series of interrelated claims that carry crucial social and political implications. In a nutshell, Gould argues that what he calls "biological determinism" (roughly, the idea that human nature sets limits on human abilities) should be rejected at all levels—scientific, political, and so forth—at least when the abilities under discussion relate in any way to human intelligence, temperament, or to the human capacity for moral or ethical action; that the concepts I have just mentioned (intelligence, temperament, and the capacity for moral or ethical action) are not in fact measurable in any scientific way and that efforts to develop measures for them represent unwarranted attempts to "reify" these concepts; and, finally, that the "whole enterprise of setting a biological value upon groups" is "irrelevant, intellectually unsound, and highly injurious."[4] Gould says that the arguments in his book are meant to discredit what he insists, not without some justification, are the currently popular theories (particularly regarding the biological basis of intelligence) that abound in these conservative times. Gould's editors at W. W. Norton and Company may well have believed this and, in the best tradition of publishing, may have tried to capitalize on the controversy surrounding these issues. It is worth noting, for example, that the front cover of the 1996 revised edition of *The Mismeasure of Man* boldly proclaims that the book is "the definitive refutation to the argument of *The Bell Curve.*"

It would be difficult to find in recent memory a more controversial popular work of social science than Richard Herrnstein and Charles Murray's 1994 book *The Bell Curve: Intelligence and Class Structure in American Life*. With all the necessary qualifications and caveats, Herrnstein and Murray still manage to argue that a type of general intelligence exists and is scientifically measurable; that it is highly heritable; that evidence supports the claim that differences we find in the average of this general intelligence among racial groups is due in some significant degree to biological differences between races; and that these biological differences in average general intelligence account (again in some significant degree) for at least some of the differences we see in income and class standing among racial groups.[5] It is also worth noting that Herrnstein and Murray insist that *their* argument is the unpopular one today, in these liberal, politically correct times. Indeed, Herrnstein and Murray open their book with an epigraph from Edmund Burke's *A Vindication of Natural Society*, in which Burke argues for the moral necessity of "a fair discussion of popular Prejudices."[6] The clear message here is that the belief that biology does not matter is simply one of *our* popular prejudices.

So the battle is joined between those who would follow Gould and those who would instead cast their lot with Herrnstein and Murray. Throughout *The Mismeasure of Man* Gould chooses to fight this battle largely in the arena of history. His argument, in a sense, is that with respect to the scientific discussion of biology and intelligence, the more things change, the more they remain the same.

Let us now return to Doctor Morton's work. Gould tells us that he spent several weeks during 1977 reanalyzing the raw data that Morton had used more than a century earlier to draw the conclusions I reported earlier about differences in cranial size among the races. By *raw data* Gould does not mean the actual skulls Morton used but rather the lists of skulls, the races of those to whom they belonged, their size measured in various ways, and so forth. It turns out that Gould could subject Morton's conclusions to interpretative scrutiny more than a century after Morton's own death precisely because Morton himself was such a meticulous data collector and organizer. Working with this raw data, Gould demonstrates that even if cranial size correlated with intelligence—and Gould argues that it does not—Morton's conclusions about differences in average cranial size among races simply are not sound. In fact, Gould argues that his own analysis of the data "reveals no significant differences among races" with respect to cranial size.[7] How could Morton have managed to misinterpret these data so drastically? Pulling no punches, Gould argues that Morton's interpretation represents nothing more than "a patchwork of fudging and finagling in the clear interest of controlling a priori convictions."[8] In other words, Morton knew what he wanted to prove and he simply massaged the data until they appeared to yield the conclusion he desired.

Some of the fudging and finagling that Gould identifies is so blatantly obvious that we must wonder how Morton himself managed to remain blind to it. For example, Gould notes that "Morton often chose to include or delete large subsamples in order to match group averages with prior expectations."[9] Hence, when a particular racial subgroup, say the Inca Peruvians, showed average skull sizes that were below the average for their racial group (North and South American Indians in general) they were included in the sample of all Indians in order to drive down the overall average. But when another racial subgroup, say Hindus, showed average skull sizes that were below the average for their racial group (Caucasians) they were excluded from the sample of all Caucasians in order to drive up the overall average. The result, of course, was to make it appear as though Caucasians on average had bigger skulls than Indians on average—precisely the claim Morton was trying to establish.

Gould notes that this little bit of fudging and finagling was not unique to Morton's work. It was common to many of his contemporaries. Nor was this the only fudging and finagling that was going on in Morton's work. Gould actually spends a considerable chunk of *The Mismeasure of Man* detailing the various ways that Morton's data simply do not fit the conclusions that were presented. But after all of the errors, omissions, and miscalculations have been identified and corrected, Gould then renders this somewhat surprising judgment on Morton:

> Through all this juggling, I detect no sign of fraud or conscious manipulation. Morton made no attempt to cover his tracks and I must presume that he was unaware he had left them. He explained all his procedures and published all his raw data. All I can discern is an a priori conviction about racial ranking so powerful that it directed his tabulations along preestablished lines. Yet Morton was widely hailed as the objectivist of his age, the man who would rescue American science from the mire of unsupported speculation.[10]

Gould's treatment of Morton's work is paradigmatic of his method throughout *The Mismeasure of Man*. That method is twofold. First, Gould reconstructs the historical and political context in which a given scientist was working. Second, he examines the raw data that the scientist was working with (as far as is possible) and demonstrates how the scientist's theories invariably tended to fit the historical and political context in which he was working better than they fit the actual data. All of this is done to illustrate what Gould calls the "cardinal principle" of his book: "the social embeddedness of science and the frequent grafting of expectation upon supposed objectivity."[11] In addition to closely examining the work of Samuel Morton, Gould subjects the work of several other prominent nineteenth- and twentieth-century scientists to careful scrutiny.

One whose work deserves special mention is Paul Broca, an extraordinarily prominent nineteenth-century French surgeon, founder of the Anthropological Society of Paris, and discoverer of an area in the frontal lobe of the brain (now known as Broca's Area) that is responsible for processing certain elements of speech. Broca was every bit as interested in brain size as was Morton, and for exactly the same reason. Broca also thought that people with large brains were smarter than people with small brains, but he was not content to infer the size of the brain from the size of the skull that contained it. He preferred instead to measure the *weight* of the brain itself, since he took weight to be the best indication of the brain's capacity for intelligence. Often he would measure the weight of a given brain after he had personally autopsied the body of the person to whom the brain belonged. Like Morton, Broca was a meticulous data collector, and during the course of his eminent career he amassed an impressive amount of data on the average brain weight of males and females and of those of various races and ethnicities. Not surprisingly, Broca's data also showed that white males had the biggest brains. Among European males, Broca's data showed, *average* brain weight varied from 1,300 grams to 1,400 grams. And, just like Morton, Broca pressed on even further, breaking down larger categories into subcategories. When Broca did this for nationalities, it just so happened that, among European males, Frenchmen had *on average* the biggest brains of all. Predictably, this ignited something of a row between French and German scientists. Even so, the largest *individual* brain that was measured belonged to the Russian novelist Ivan Turgenev, who weighed in (astonishingly) at over 2,000 grams. The French and the Germans were still well represented, however. The brain of the famous French anatomist Georges Cuvier weighted 1,830 grams, putting him well within the genius category, while the brain of J. K. Spurzheim, a German and one of the founders of phrenology, weighed 1,559 grams, well above average if not quite in the genius realm. Interestingly, Gould reports that "the largest female brain ever weighed (1,565 grams) belonged to a woman who had killed her husband."[12]

There is perhaps something mildly entertaining about cataloging the brain weights of famous people, and I could go on doing so. But anyone can see where this is ultimately headed. Just as in the case of Morton, Gould demonstrates that Broca massaged his data in order to obtain the results he sought, often in ways that seem in retrospect comically capricious. In particular, on those not infrequent occasions when Broca came across the brain of an otherwise intelligent person that just did not happen to weigh enough, he would simply adjust the weight upward and justify the addition by claiming that the actual brain had shrunk from its "normal" size either because of the advanced age of the person (brains do shrink with age) or because of the person's diet (which can be a factor) or because of the person's poor overall health or because of the way the brain was preserved before measurement. When all else failed he would simply deny that the person was really as intelligent as everyone had initially assumed. Such, it seems, was the fate of several eminent German university professors. After their brains came in at or below average weight, Broca decided that being a professor did not automatically mean that one was intelligent. When even this bit of finagling failed him, Broca would simply ignore the brain or treat it as an unexplainable anomaly. That is apparently what happened in the case of Karl Friedrich von Gauss, one of the greatest mathematicians in history and undeniably an individual of extraordinary genius. After careful measurement, Gauss's brain was found to weigh only 1,492 grams, slightly above average for a European male, but definitely not in the genius range, at least according to Broca's criterion. Finally, I suppose it would be a bit unkind of me to conclude this discussion without answering the question that I imagine many readers have on their minds. Without comment, then, and for what it is worth: Broca's own brain weighed 1,424 grams.

The point I want to emphasize is the similarity between Morton and Broca. Both were extremely well respected men of science; both were committed to "objectivity" in science. As far as we know (and we seem to know a good deal) neither was a fraud, in the sense that neither lied about or even misrepresented his raw data. Indeed, both men seem to have gone out of their way to make their data and conclusions as open as possible to scientific scrutiny. Yet, as Gould deftly demonstrates, the conclusions reached by each man, particularly about race and intelligence, seem to be dictated not by the data but by biases of which neither man may well have been consciously aware.

Keeping that point in mind, I want now to quote Gould's general assessment of Paul Broca. While the passage I shall quote mentions Broca specifically, it can and should be generalized to represent Gould's thinking about the enterprise of science itself. Indeed, the following quotation is my candidate for the single most important passage in Gould's book. I give it that honor largely because it states so well what Gould has already called the cardinal principle of *The Mismeasure of Man*. The key point Gould wants us to understand is this:

> Science is rooted in creative interpretation. Numbers suggest, constrain, and refute; they do not, by themselves, specify the content of scientific theories. Theories are built upon the interpretations of numbers, and interpreters are often trapped by their own rhetoric. They believe in their own objectivity, and fail to discern the prejudice that leads them to one interpretation among many consistent with their numbers. Paul Broca is now distant enough. We can stand back and show that he used numbers not to generate new theories but to illustrate a priori conclusions. Shall we believe that science is different today simply because we share the cultural context of most practicing scientists and mistake its influence for objective truth? Broca was an exemplary scientist; no one has ever surpassed him in meticulous care and accuracy of measurement. *By what right, other than our own biases, can we*

identify his prejudice and hold that science now operates independently of culture and class?[13]

There are at least two possible interpretations of the above passage: the weak interpretation and the strong interpretation. The weak interpretation takes Gould as making the utterly obvious and uncontroversial point that scientists, like everyone else, are human, all too human, and hence subject to all the conscious and unconscious biases of the rest of us. No one disputes this, least of all practicing scientists. But this is, I will concede, a point that cannot be made often enough.

The strong interpretation takes Gould as saying that scientists cannot, *even in theory*, step outside of their particular culture or class, precisely because there is (in reality?) no "reality" outside of one's own culture or class. This interpretation historicizes science completely. Gould practically dares us to think ourselves superior to the great scientists of the past. He is confident that we will shrink from this challenge once we understand how right these past scientists thought they were, but how wrong we now *know* them to be. The safest course, it seems, is to believe that science, like everything else, is a mere product of a particular historically situated culture or class. Gould, I take it, sincerely believed that this course frees us from the oppressive biological determinism of *our* culture. It allows us to think that biology is not destiny and that human equality is possible after all.

The problem, however, is that the course Gould charts for us takes us to the brink of the self-referential paradox. For if science is merely a product of the culture or class of the individual scientist then Gould's own culture or class, and *not* his science, becomes the focus of any discussion of *The Mismeasure of Man*. Perhaps, then, this is the appropriate time to note that for all of his adult life Stephen Jay Gould was (intellectually, at least) a Marxist whose own cultural and class position, as an intellectual and a member of Harvard's faculty, made it extremely likely that he would seek to advance the liberal left agenda. During the late twentieth century that agenda seemed to consist in calls for greater federal government involvement in the economy as a way of shoring up the welfare state and calls for greater enforcement of policies that would promote the equality of results as opposed to the equality of opportunity, all coupled with very loud calls for more "democracy." These calls for more democracy served the rhetorical function of deflecting attention from the fact that the type of egalitarian-socialist society envisioned by the left would leave precious little room for real democracy, insofar as the state would need to be micromanaged by a cadre of intellectual elites—the very kind of people, in fact, who comprise the Harvard faculty. Such, it seems, was Gould's ideological prejudice. It is a prejudice all too apparent in *The Mismeasure of Man*, especially in the introduction to the revised and expanded edition, in which Gould forcefully attacks the allegedly elitist and anti-democratic agenda of Herrnstein and Murray's *The Bell Curve*.

But at this point, anyone on the above-average side of the bell curve will surely notice the potentially paradoxical, if not self-contradictory, nature of Gould's argument. If science is simply the reflection of the cultural or class prejudices of scientists, then it is not clear that there is any nonprejudicial reason for believing the claims advanced by Gould (claims, for example, about the nonexistence of any measure of general intelligence) over the exactly opposite claims of Herrnstein and Murray (claims, for example, about the existence of a general measure of intelligence and about its high heritability). Gould, it seems, cannot without contradicting himself argue that Herrnstein and Murray are *scientifically* wrong when they say that a general intelligence exists and that it differs among races. All Gould can argue is that the prejudices of the authors of *The Bell Curve* are not his prejudices. To be sure, this is a powerful argument in a culture in which egalitarianism is a very popular prejudice. But ultimately its power is fragile and fleeting. Ours is a culture that demands its ideas and principles be grounded in something more than mere cultural or class prejudice. Gould doubtless understood this, and he surely knew that *The Mismeasure of Man* would lose all of its rhetorical force if he conceded outright that all of science was simply the mere reflection of cultural or class prejudice. In the face of such a concession, why would anyone struggle through the sometimes complicated statistical analysis Gould provides at the end of his book—analysis that Gould claims disproves the assertions about intelligence made by Charles Spearman and Cyril Burt, the early-twentieth-century predecessors of Herrnstein and Murray? Indeed, why would anyone bother to read Gould at all, when one could simply pick up at random a book by some third-rate science writer or ideological hack that made the same point as *The Mismeasure of Man* but that demanded much less of its reader? Gould knew that it matters the way one invalidates a scientific theory or claim. Thus, although he may have taken us to the brink of the self-referential paradox in *The Mismeasure of Man*, he is careful not to go over the brink. Or, to change metaphor, one might say that Gould inoculated himself against the charge of self-referentiality in the introduction to his book. He begins by arguing that "the most creative theories [in science] are often imaginative visions imposed upon facts; the source of imagination is also strongly cultural." But he then quickly adds,

> This argument, although still anathema to many practicing scientists, would, I think, be accepted by nearly every historian of science. In advancing it, however, I do not ally myself with an overextension now popular in some historical circles: the purely relativistic claim that scientific change only reflects the modification of social contexts, that truth is a meaningless notion outside cultural assumptions, and that science can therefore provide no enduring answers. As a practicing scientist, I share the credo of my colleagues: I believe that a factual reality exists and that science, though often in an obtuse and erratic manner, can learn about it.[14]

There is, I would argue, at least something of a tension between this passage and the one quoted earlier challenging our "right" to hold that science now operates independently of culture and class. One wonders if, to his doctrinaire Marxist colleagues, Gould's embrace of the scientific credo mentioned above might represent simply a failure of nerve on his part. These Marxists might well ask by what right, other than his own prejudices as a practicing scientist, Gould could take the cultural context of most practicing scientists *of his day* as opening up even the possibility for objective truth.

The point, it seems, is that one cannot serve two masters. In the final analysis, in order to be a scientist one must choose to believe (as Gould apparently did) that cultural, historical, or political knowledge, though valuable in its own right, is not particularly helpful in evaluating the truth of scientific theories *as such*. Indeed, the historical lesson I would like to take away from this discussion of *The Mismeasure of Man* and apply to evolutionary psychology is as follows: Even if evolutionary psychology represents yet another attempt to develop a modern science of man, and even if past attempts to do the same now seem like abject failures, that historical knowledge is still not sufficient to dismiss evolutionary psychology itself.

But surely historical knowledge is not unimportant. With that in mind, in the next three chapters I discuss some of the most important and interesting battles that have been fought during the course of the last three decades in the ongoing nature-versus-nurture debate. My intention is to provide a sketch of the recent social, political, and scientific context within which evolutionary psychology developed. I begin with a discussion of some very controversial ideas about the nature of human intelligence that were publicly advanced in 1971 by Richard Herrnstein, then a professor of psychology at Harvard University. Chapter three then examines the ideas of another Harvard professor, Edward O. Wilson. As the author of *Sociobiology: The New Synthesis*, Wilson may rightly be regarded as the intellectual father of evolutionary psychology. While I do not intend to provide a full history of the development of sociobiology, I do hope to provide the reader with both an understanding of its central principles and a feeling for the controversy it generated.[15] Finally, chapter four offers an examination of the major critical response to sociobiology that was published in 1984 by three distinguished scientists who attempt finally to settle the nature-versus-nurture debate.

CHAPTER TWO

Richard Herrnstein Stirs Up
Controversy at Harvard Yard

Less than two months before *The Bell Curve* was published in October of 1994, one of its coauthors, Richard Herrnstein, died. Herrnstein was thus spared the ordeal of having to endure the firestorm of controversy that he helped to create with that book. But Herrnstein was no stranger to controversy. In fact, the central argument of *The Bell Curve* was advanced twenty-three years before the book's publication in a lengthy article by Herrnstein entitled simply "I.Q." The article, which appeared in the September 1971 issue of the *Atlantic Monthly*, is interesting not just as a scientific, but also as a *rhetorical*, document.[1] By writing in the *Atlantic Monthly*, one of this country's few popular magazines devoted to examinations of public issues on a broad but fairly high intellectual level, Herrnstein clearly hoped to reach a general audience of educated Americans who might be receptive to some reasonably fresh (or at least freshly formulated) ideas regarding the perennial nature-versus-nurture debate, especially as it applied to the issue of human intelligence.

Herrnstein seems to have wanted to do at least three important things with his article. First, he wanted to convince his audience that IQ tests really did work at measuring something that could reasonably be called general intelligence—something that was different from mere cultural knowledge. Second, he wanted to introduce to his audience, and to defend, the work of Arthur Jensen, a professor at the University of California, Berkeley, whose views on intelligence were at the time probably more controversial even than those of Herrnstein himself. And third, Herrnstein wanted to advance and defend the ironic but critically important claim that the more equality of opportunity a society provides, the more unequal the society has the potential to become.

Whatever else he did, Herrnstein accomplished with his article the initial goal of anyone who seeks to influence public debate on an issue. He got noticed. As Ullica Segerstråle reports in her book *Defenders of the Truth: The Battle for Science in the Sociobiology Debate and Beyond*, in the years following the appearance of Herrnstein's article, "his lectures were interrupted, and posters around Harvard yard pictured him as 'wanted' for racism."[2] While this reaction may not be representative of the overall response to Herrnstein's article by most of his readers, there is no question that Herrnstein angered many people. But what angers a person or a culture tells us more about that person or culture than perhaps any other piece of information. Thus it is worth examining exactly what Herrnstein was doing in his article in order to see precisely why it was so controversial.

The first thing Herrnstein does is advance the dual claim that intelligence can in fact be measured and that what is being measured as intelligence is more than simply book-learning or cultural knowledge. He begins by noting the interesting connection between the study of human intelligence and the study of evolution. It turns out that the Darwinian family tree was central to both studies. Many people know that in 1859 Charles Darwin published *The Origin of Species*, thereby inaugurating the modern study of evolution. But exactly a decade later, Charles's younger cousin, Francis Galton, published *Hereditary Genius*, a book whose title presents its central argument: that intelligence is hereditary and thus genius runs in families. There is some indication that Galton got the idea for his book from his (unbiased?) observation that genius seemed to run in his own family. There is no question that Galton hoped to use the insights in *Hereditary Genius* for political and social ends. He coined the term *eugenics* and firmly believed that the human race could be improved through selective breeding.[3]

The big problem with Galton's research, however, was that it begged the absolutely critical question of what, exactly, constituted intelligence. One would think that without a precise answer to this question the measurement of intelligence would be impossible. But a precise answer was very difficult to obtain. Everyone agreed that intelligence was more complicated than height or weight, which could be observed and measured directly with extremely simple instruments. But beyond that no one was quite certain what he was talking about, or even whether he was talking about the same "thing" as his colleagues who also happened to be researching intelligence. Galton and other nineteenth-century researchers on intelligence knew that, to some degree at least, people's differing abilities to solve daily problems and to cope with the world on various levels were in some way related to some "thing" going on in people's heads. The problem was measuring, or even defining, this "thing."

One possible solution to the problem of defining intelligence—or more accurately one possible way *around* the problem—was offered in the early twentieth century by Alfred Binet, a French psychologist whose name has become synonymous with IQ tests. As Herrnstein explains, Binet's contribution to the study and measurement of intelligence lay in his insight that while intelligence might be difficult to define in a strict sense, it could be measured as a relative difference within a given population. Binet worked extensively with children. He noted that most children in a given age group have about the same knowledge of their world and do very nearly the same on skills like copying shapes on a piece of paper or simply counting objects. Some children, however, do much better at these activities than others in their age group and some children do much worse. Binet used this information to develop the first modern IQ test. When given to large numbers of children, Binet's test seemed to confirm what many teachers seemed to have noticed: on average children who were deemed "smart" tended to possess the types of skills and knowledge possessed by average children older than themselves while on average children who were deemed "dull" tended to possess skills and knowledge possessed by average children younger than themselves.

But still the question might be asked, What exactly are we measuring here? Could it not be that a given child who happened to do well on Binet's test and who happened to be labeled smart by her teacher was not in fact smart, but was only *privileged*, in the sense that she had access to important information or training that other children who tested poorly or were labeled dull did not have? This takes us directly to the issue of *cultural bias*, an issue that has bedeviled intelligence researchers at least since the days of Binet. As it turned out, some of the questions on a 1911 version of one of Binet's tests do seem biased, in the sense that a person's cultural knowledge might aid him or her in answering these questions. For example, an average six-year-old child was expected to be able to "define familiar objects in terms of use," and an average ten-year-old child was expected to be able to "criticize absurd statements." It will immediately be objected that what is *familiar* or *absurd* to a child may well depend considerably on that child's cultural background. But other questions seem less likely to have drawn on specific cultural knowledge. For example, an average six-year-old child was also expected to be able to "copy a diamond shape" and "count thirteen pennies," while an average ten-year-old child was expected to be able to "arrange five blocks in order of weight" and "draw two designs from memory."[4] To be sure, even these last four skills might depend to some degree on a child's cultural background (if, for example, the design that the child was expected to draw from memory happened to be more familiar in one culture than another). But Binet realized this. Hence he consciously sought to design his tests so that the majority of children in any given age group did about the same, while about equal proportions did well and poorly. More specifically, he designed his test so that the results for a sufficiently large sample would be normally distributed around a mean—in other words, so that the results would plot out a bell curve. Binet seems to have been relatively successful at this. Herrnstein sums up Binet's test with this general assessment: "It was not only in France that the average eight-year-old child could just barely repeat accurately five digits read to him, for the Binet scale was readily exported to Belgium, Great Britain, America, Italy, and so on. *The remarkable exportability of the tests was probably the first convincing argument for their soundness.*"[5]

Although Binet worked largely with children, this does not invalidate his central insight that any given individual's intelligence can be measured relative to a given population. Thus today, when speaking of *adult* intelligence, we tend not to use terms like *mental age*, unless the individual in question is at the extreme low end of the intelligence scale, in which case we may say (for example) that he or she has the mental age of a ten-year-old child. For all other adults we tend to say that a given individual's intelligence places him or her in (for example) the upper 10 percent of the (adult) population.

The critical point to see in our discussion thus far is that—as Herrnstein emphasizes—"Binet invented the modern intelligence test without saying what intelligence is."[6] Certainly Binet had no idea of what neurological changes were taking place inside a person's brain as he or she attempted to draw a figure or count pennies. All Binet had were numerical scores on various tests administered to a large number of individuals. From this he inferred that the differences in these test scores were best explained by the differences in the levels of intelligence of the individuals taking the tests. But here we encounter another question about the measurement of intelligence: Even if Binet's tests displayed no significant bias in favor of one culture or another, how do we know that his test measured a given individual's intelligence, as opposed to simply measuring . . . well, measuring a given individual's ability to answer the particular questions on the test? After all, the only thing we can say for certain about an individual who correctly answers every question on the Scholastic Assessment Test is that she can correctly answer all the questions on that particular test. (And even this is not certain, for she might not have been able to answer all the questions and therefore guessed on some.)

A consideration of this question brings us to the second part of the first claim that Herrnstein makes in his article, and to a discussion of the existence of a *general intelligence*. The idea that the intelligence tests administered by Binet and his colleagues, with their various types of questions, may have been measuring one specific, and apparently deep, intellectual ability was not immediately obvious to Binet himself. As Herrnstein points out, Binet almost certainly did not believe that intelligence was one specific "thing." Still, Binet did not make this question central to his own research. The first individual to do so with any success was Charles Spearman, a former British army officer and "fine statistician," according to Gould, who is here damning with faint praise.[7] In 1904 Spearman published a paper, "General Intelligence Objectively Measured and Determined," the title of which seems both self-explanatory and definitive. As we shall see, however, not everyone agrees with Spearman's conclusion that something like a general intelligence exists and can be measured. But Herrnstein clearly did. And although Spearman himself gets only a relatively brief mention in Herrnstein's *Atlantic Monthly* article, the concept of general intelligence plays a central role in its thesis. This is how Herrnstein introduces the concept: "Taking the intercorrelations between scores on simple mental tests as his basis, [Spearman] concluded that there was a 'universal' intellectual capacity—which he labeled 'g' for 'general'—plus a host of minor, unrelated capacities of no great scope. The universal factor, he said, permeated all intellectual activity, while the others were variously absent or present in any given task. To be smart, for Spearman, mainly meant having lots of *g*."[8]

I admit that when I first read about this *g*, I was skeptical. In 1994, when *The Bell Curve* first came out, I quickly read through the introduction in a bookstore. The concept of general intelligence is also central to the argument of *The Bell Curve*. On page three of the introduction, Herrnstein and Murray note that Spearman "uncovered evidence for a unitary mental factor, which he named *g*, for 'general intelligence,'" and they briefly recount the evidence that led to this discovery:

Spearman noted that as the data from many different mental tests were accumulating, a curious result kept turning up: If the same group of people took two different mental tests, anyone who did well (or poorly) on one test tended to do similarly well (or poorly) on the other. In statistical terms, the scores on the two tests were positively correlated. This outcome did not seem to depend on the specific content of the tests. As long as the tests involved cognitive skills of one sort or another, the positive correlations appeared.[9]

As I said, I was at first skeptical. I would bet my reaction was not atypical. In our personal lives we all know (or think we know) someone who is a computer genius but who cannot write a coherent sentence, or someone who is a gifted writer but who cannot balance a checkbook. Does not the existence of such individuals disprove the existence of a general intelligence? The short answer is that it would, if such individuals did in fact exist in any significant number. But they do not. A more detailed answer would point out that the existence of *g* does not imply that on average a person who does well on one mental test will do *exactly* as well on another. Thus the existence of *g* would not be disproved if we tended to find that individuals who recorded a perfect score on the quantitative section of the Scholastic Assessment Test did not also record a perfect score on the verbal section of that test. Nor would the existence of *g* be disproved even if we found that individuals who recorded a perfect score on the quantitative section of the SAT recorded only an above average score on the verbal section. On the other hand, the existence of *g* would need to be seriously questioned if we found that, for a sufficiently large sample, individuals who tended to score above average on the quantitative section of the SAT tended to score below average on the verbal section. This we do not find.

I have already noted that the existence of *g* plays a critical role both in Herrnstein's *Atlantic Monthly* article and in *The Bell Curve*. But what kind of "thing" might this general intelligence be? In *The Bell Curve*, Herrnstein and Murray, following the work of Arthur Jensen (whom we will meet shortly), cannot resist using a computer analogy to explain *g*. Hence they hypothesize that *g* might in some ways be analogous to "the microprocessor in a computer."[10] This analogy suggests that very smart people may be working with an internal processor like the Intel Pentium 4 (one of the fastest commercial chips available as I write this in 2002) while average people may be working with an Intel Celeron (something like what one finds in the average computer today), and very dull people may be working with only an Intel

8088 (the chip that went into some of the very first personal computers developed in the early 1980s).

It is just this type of analogy that tends to enrage those like Gould who are critical of the concept of a general intelligence. In fact, Gould makes a discussion of general intelligence the focus of the last third of *The Mismeasure of Man*. In keeping with his method throughout, Gould closely examines Spearman's work and finds that it reveals a high degree of numerical finagling designed to prove the desired hypothesis: that general intelligence does in fact exist. In order to do any justice at all to Gould's criticism of Spearman's *g* I must now beg the reader's indulgence for a very lengthy but necessary digression about the meaning of the term *correlation* in statistics and about the idea of *causality* in the social sciences. I can justify this digression somewhat by noting that it will not only aid in the discussion of Spearman but that it will also be of much use in chapter four when I discuss yet more criticisms of recent attempts to put intelligence more firmly on a biological foundation.

To begin our digression, imagine that you find yourself in a casino. This is as good a place as any, and maybe a better place than most, to study and observe statistical relationships of all sorts. You start by wandering over to the blackjack table, and you notice that, for most male players, the greater the amount of an individual's average bet per hand the greater the number of women that are within five feet of him. There may be a *correlation* between the amount of a male's bet and the number of women he attracts. A correlation is defined as the measure of similarity between two quantities that can vary. That is, a correlation is the degree of similarity between two variables. A perfect positive correlation is expressed by the numeral 1; no correlation at all is expressed by the numeral zero; and a perfect negative (or *inverse*) correlation is expressed by the numeral −1. Of course, almost nothing in the world is perfect. In the above example, we seem to have a positive correlation between the size of a male's bet and the number of women around him. But you might observe one or two males making relatively small bets yet attracting large numbers of women. You might then want to explore further possible explanations. You may want to look for stronger correlations. Perhaps it is not the size of a male's bet that is attracting women, but rather the stylishness of his clothing. Assuming that you could *quantify* the variable *stylishness*—by, for example, reducing it to the price of a given male's clothes or the number of recognized designer items that a given male is wearing—you could test to see whether stylishness was positively (or negatively) correlated with attractiveness to women or whether there was no correlation at all. By the way, you would almost certainly notice that stylishness was positively correlated with the size of a male's bet, especially if you use the price of clothing as the measure of stylishness. By further refining your experiments and observations you might be able to discover some very strong correlations between the size of a male's bet and several variables. This would give you good information about patterns or regularities in a small slice of society. Finding patterns or regularities is of course the raison d'être of science in general, and particularly of the discipline of statistics as it is used throughout the social sciences.

Now, without leaving the casino, suppose you wander from the blackjack table to the slot machines. You notice that, for every one hundred coins played in any given slot machine, the number of times that particular slot machine pays off *decreases* as the distance of the slot machine from any entrance to the casino *increases*. In this case, you may be noticing a negative correlation between the number of times a machine pays off and its distance from a casino entrance. The *greater* the distance the *fewer* the number of pay offs per one hundred tries. Incidentally, if the casino owner is smart this correlation may hold for the number of pay offs per one hundred coins played, but *not* for the total dollar amount of all the pay offs per one hundred coins played.

Leaving the slot machines, you now wander over to the roulette table. You observe an obsessive individual who plays only one number, black 13. The size of this individual's wager does, however, vary. On one spin he might bet a little, on another spin he might bet a lot. You are curious about the correlation, if any, between the size of this individual's wager per spin of the wheel and the likelihood that the little white ball will land in the slot marked black 13. Assuming that everything is on the up and up, you should find no correlation at all—that is, you should find a correlation of zero. The reason is that the *likelihood* that the little white ball will land in the slot marked black 13 *does not vary*. In most American casinos it is always 1 in 38, or about 2.6 percent. Remember, a correlation (either positive or negative) can exist only between two *variables*. Interestingly, if you noticed that there was a *positive* correlation between the amount of this obsessive roulette player's wager and the likelihood of a pay off, you might well suspect the individual of cheating by somehow manipulating the wheel to his favor. On the other hand, if you noticed there was a *negative* correlation between the amount of the individual's wager and the likelihood of a pay off, you might suspect the croupier of cheating.

Before leaving the casino (and this digression) behind I need to discuss briefly the difference between the concept of *correlation* and the concept of *causation*. Suppose, finally, that as you walk throughout the casino you notice that there is a positive correlation between the amount of alcohol that a person drinks while in the casino and the amount of money the individual wagers. You may infer from this correlation that the more an individual drinks the more addled his or her judgment becomes, hence the more likely he or she is to make a large wager, even in the face of the inescapable fact that *in the long run* the players, not the casino, always lose. If you did infer this, you would be inferring a causality from your observed correlation. You would be inferring that increased drinking *causes* increased wagering. But this inference about causation might be false, even if the correlation is true. It might be the case that alcohol does not impair anyone's judgment, but it does help everyone drown his or her sorrows. Since, in the long run, the more a person bets the more he or she loses, the greater will be his or her sorrows, and hence the more he or she will want to drink. Thus, it may be the case that increased wagering *causes* increased drinking. Or it may be the case that neither causal explanation accurately accounts for the observed correlation. It may be the case that some third factor is responsible. Certain naturally occurring chemicals in one's brain may predispose one to drink *and* to gamble—perhaps because both are addictive, or "thrilling," behaviors. Or it may be the case that still another factor—like age, intelligence, or socioeconomic status—is responsible for the initial correlation you observed.

It turns out that *nothing* bedevils the social sciences more than the concept of causality. This is because correlations are easy to establish, but causality is not. Unfortunately, causality is always ultimately what is at issue. Suppose, for example, that you sought to reduce gambling by passing a law that prohibited individuals from drinking alcoholic beverages while in casinos. As a legislator, you may have noticed a positive correlation between drinking and gambling. But if that correlation *were* caused largely by a natural variation in a particular brain chemical, then your law would be largely ineffective. Presumably you would notice this after the law was passed and the level of wagering at casinos remained unchanged.

But where matters of public policy are concerned, we would like to know what causes what *before* we start passing laws. Hence we turn to our social scientists, who are forever doing studies to find correlations and, ultimately, to establish causality. Many of these studies are highly controversial, none more so than those involving human intelligence. We would do well to keep this in mind as we return from what has been a fairly lengthy, but I hope useful, digression. Before the digression, we were examining Gould's criticism of Charles Spearman's argument for the existence of a general intelligence, or g. Spearman inferred the existence of this g from high positive correlations he noticed on the test scores of individuals who had taken various intelligence tests. In fact, as Herrnstein and Murray point out in *The Bell Curve*, it is "nearly impossible" to design a test that measures anything like intelligence and that is *not* positively correlated with other tests that measure anything like intelligence.[11]

Gould does not dispute the evidence showing high positive correlations between any number of tests that seem to measure intelligence. He does, however, vigorously dispute the *interpretation* of this evidence. He insists that, although these correlations exist, that fact alone is not proof that the correlations in question spring from the existence of some general intelligence. According to Gould, there is no reason to assume that these correlations spring from any "thing." They may just be statistical artifacts. As Gould points out, it is not difficult to generate a positive (or negative) correlation. All you really need are any two variables. Thus, since over the course of the last five years Halley's comet has been moving away from the earth, we notice that there is a positive correlation between the distance of Halley's comet from the earth and the number of websites on the world wide web within the last five years.[12] Simply put, both variables have been increasing over the last five years.[13] But it would be absurd to suppose that this correlation were caused by any "thing," much less by some deep force in the universe that affected comets and websites.

Gould also emphasizes that even when we do observe what appears to be a meaningful correlation between variables, it is often impossible to say what is causing the correlation. The correlations we observe between test scores for intelligence make this point perfectly, according to Gould. We *could* say, he concedes, that these correlations were caused by some innate general intelligence. But we could with equal confidence say, Gould insists, that these correlations were caused by the usual standbys: education and environment. It may well be that all the measures we are using to test for intelligence stem from the same educational biases or intellectual environment, and that these biases and this environment also happen to form a large part of the life of people who happen to do well on intelligence tests.

Finally, Gould insists that the principal error in Spearman's work is that he is guilty of reification. In the context of research on intelligence, reification involves "the notion that such a nebulous, socially defined concept as intelligence might be identified as a 'thing' with a locus in the brain and a definite degree of heritability—and that it might be measured as a single number, thus permitting a unilinear ranking of people according to the amount of it they possess."[14] In a deft rhetorical move, Gould argues that social scientists who tend to reify the concepts with which they work may well be suffering from a form of "physics envy." They want their research to be taken as seriously as the "hard sciences," so they invent real "things" to study. This way they are not forced to say that they work only with statistical correlations.[15]

These are just some of the problems that Gould finds in Spearman's research. He actually presents a much more detailed critique of Spearman's concept of *g* than this summary suggests. But throughout his critique Gould continuously presses the point that intelligence researchers like Spearman, Galton, Cyril Burt, L. L. Thurstone, and others just placed too much faith in numbers. To be sure, Gould is correct that these men did in fact place a great deal of faith in numbers and, specifically, in the branch of mathematics that deals with probability and statistics. Indeed, all of the individuals just mentioned not only relied heavily on statistics in their work but also contributed genuine mathematical advances to that field.

But was their faith in numbers really so unwarranted? In attempting to arrive at a balanced answer to this question we should not forget that the laws of statistics do provide us with some profoundly powerful tools with which to predict and control our world. Consider, for example, one of the most important laws in statistics—variously known as the central limit theorem, or the normal law of error, or the law of frequency of error. This law says that as you increase the size of a simple random sample drawn from almost any population (regardless of how skewed the actual population may be) the values for the mean of your sample will tend to become normally distributed around the mean of the actual population.[16] This law is precisely what allows researchers to make valid statistical inferences from (actually fairly small) random samples of very large populations. To put the point more concretely, it is just this law that allows a pollster to announce (with 95 percent confidence, as statisticians say) that a given candidate will receive 40 percent (plus or minus, say, 4 percent) of the number of votes cast in an election, even though the pollster has randomly sampled only a small but statistically significant percentage of all voters. The more clearly you understand this law, the easier it is to perhaps become caught up in an appreciation of its power. This may account for the following quotation from William Youden, a twentieth-century research chemist turned statistician who became an apostle for the power of numbers: "The normal law of error stands out in the experiences of mankind as one of the broadest generalizations of natural philosophy. It serves as the guiding instrument in researches in the physical and social sciences and in medicine, agriculture, and engineering. It is an indispensable tool for the analysis and the interpretation of the basic data obtained by observation and experimentation."[17]

Referring to the same law as Youden, only much more poetically, Francis Galton wrote, "I know of scarcely anything so apt to impress the imagination as the wonderful form of cosmic order expressed by the 'Law of Frequency of Error.' The law would have been personified by the Greeks and deified, if they had known it."[18] It is precisely this kind of talk that seems to have upset Gould, especially in relation to Spearman's conception of general intelligence. Gould wants us to believe that general intelligence is not real, but merely a property of the clever manipulation of some numbers. Because the debate over the concept of general intelligence is so important, I cannot leave this discussion without mentioning one of the responses that Charles Murray provides to Gould's arguments against the existence of *g*. Murray's response appears in a 1996 afterword to *The Bell Curve*. After reviewing many of the criticisms Gould presents in *The Mismeasure of Man* against Spearman's concept of *g*, Murray insists nonetheless that this concept does indeed "capture a 'real property in the head.'" To support this claim, Murray reviews some evidence from the field of neuroscience. Before quoting Murray's analysis of this evidence I need to explain the meaning of *loading* as that term refers to *g*.

To say that a given intelligence test (or even part of that test) is highly *g* loaded, means that the scores on the test (or part of the test) in question correlate very highly with the scores on other intelligence tests (which themselves correlate highly one with another). To put it another way, to say that one test is more highly *g* loaded than another is to say that the one test *draws more heavily* on *g* than the other. To use a simple but elegant example that Herrnstein and Murray also use, consider a test that measured intelligence by assessing a person's ability to repeat a series of numbers read to him or her. Such tests correlate positively with the intercorrelations of other tests of intelligence. In other words, they measure *g* to some extent. But now change the test slightly by asking the person to repeat *in reverse order* a series of numbers read to him or her. It turns out that a test that assesses this second ability is more highly g loaded than a test that simply assesses a person's ability to repeat numbers in normal order.

With this in mind, we are in a better position to understand Murray's claim that g captures something real in the head. Murray notes that

a growing body of evidence links g, and IQ scores more generally, with neurophysiological functioning and a genetic ground: The higher the g load ing of a subtest is, the higher is its heritability. The higher the g loading of a subtest is, the higher is the degree of inbreeding depression (an established genetic phenomenon). Reaction times on elementary cognitive tasks that require no conscious thought, such as responding to a lighted button, show a significant correlation with IQ test scores. This correlation depends mostly, perhaps entirely, on g. A significant relationship exists between g and evoked electrical potentials of the cerebral cortex. A significant inverse relationship exists between nonverbal (and highly g loaded) IQ test scores and the brain's consumption of glucose in the areas of the brain tapped by the cognitive test. The higher the scores are on IQ tests, the faster is the speed of neural and synaptic transmission in the visual tract.[19]

From this Murray concludes that "the reality and importance of g has long since, in many ways, been established *independent of its statistical properties.*"[20] Finally, Murray draws our attention to an open letter concerning "main-stream science on intelligence" that was published in 1994 in the *Wall Street Journal.* The letter was signed by fifty-two professors, "all experts in intelligence and allied fields." As Murray correctly points out, the letter endorses "all of the main scientific findings of *The Bell Curve,*" including the existence and measurability of a general intelligence.[21] Thus, as Murray says, "the big news about the study of intelligence is not that science has moved beyond the concept of a general mental ability but the remarkable resilience and utility of this construct called g."[22]

It seems that the only certain thing about the debate over the existence of a general intelligence is that the debate itself has been going on for nearly one hundred years and that it still generates a good deal of controversy. But why is it so important to know whether this g exists? One answer is that proof of the existence of g would seem to put research into intelligence on firmer ground. Researchers would know they were measuring something more than a statistical fluke. Another reason that it might be important to confirm the existence of g involves the types of challenges (particularly occupational challenges) people will be facing as society continues to advance. The point to be made here is central both to Herrnstein's original *Atlantic Monthly* article and to *The Bell Curve*. The point is that as society becomes more advanced, the types of jobs people perform in the society will tend to change very rapidly; hence, people may need to be retrained for various jobs throughout the course of their working lives. At the very least, to remain employed in advanced industrial societies individuals must continuously learn new things, and perhaps acquire new skills, even if they do not formally change jobs or careers. If general intelligence existed and could be measured, it might give us an especially relevant look at how well a given individual would be likely to do in an advanced industrial society.

Thus the debate about the existence and measurability of general intelligence is more than merely academic. It could have significant public policy implications. Again, if *g* existed and could be measured, it might lend support to the use of general aptitude or intelligence tests instead of more specific content-based tests. (Interestingly, what is now called the Scholastic Assessment Test used to be called the Scholastic Aptitude Test, precisely because its general questions are designed to measure a high school student's aptitude for college work, not his or her knowledge of any given subject. The name of the test has changed—for politically correct reasons the word *aptitude* was deemed offensive—but the rationale for the test has remained the same.) Also, if *g* existed and could be measured (perhaps by SAT tests), it might serve as a means of tracking those with high *g* scores into broader, more general educational environments (like four-year liberal arts colleges) while tracking those with low *g* scores into more specific vocational programs. Of course, if one had moral or ethical objections to such tracking, then proof that *g* did *not* exist and that it was merely a clever statistical contrivance of the wealthy and the powerful would be welcome news indeed.

At any rate, the debate about the existence of a general intelligence, its measurability, and its potential impact on public policy questions continues to this day. Herrnstein's original *Atlantic Monthly* article both anticipated and fueled much of this debate. To sum up: the first claim that Herrnstein develops and defends in his 1971 article is that intelligence exists and that it is measurable. Herrnstein develops this argument by examining the work of Francis Galton, Alfred Binet, Charles Spearman, and others. He also pays particularly close attention to the idea that there is a general intelligence that is to greater or lesser degrees measured by all types of intelligence tests.

Had Herrnstein stopped there, his article would have been a little shorter and probably a lot less controversial. But Herrnstein's purpose was not simply to make a point about intelligence. It was also to make a point about science and society. Thus he goes on to discuss and defend the work of Arthur Jensen, a professor at the University of California, Berkeley. As I have said, at the time Herrnstein's article appeared, Jensen was a more controversial figure in the field of intelligence research than Herrnstein himself. Indeed, the first mention we get of Jensen in Herrnstein's article comes by way of a reference to Jensen's infamous 1969 article published in the *Harvard Educational Review* and entitled "How Much Can We Boost I.Q. and Scholastic Achievement?"[23] Jensen's simple and direct answer to that question is, *not much, if at all.* But it is important to understand precisely what Jensen is and is not saying. He is *not* saying that improved nutrition and health care, and the eradication of environmental hazards like lead-based paints, are not likely to improve the overall general intelligence of children whose scholastic performance is below average. He *is* saying that *remedial education* will not measurably improve it.

Herrnstein actually quotes several paragraphs of Jensen's article. But the very first sentence of the article sets the tone unmistakably. Jensen writes, "Compensatory education has been tried and it apparently has failed."[24] He then goes on to argue that "the chief goal of compensatory education—to remedy the educational lag of disadvantaged children and thereby narrow the achievement gap between 'minority' and 'majority' pupils—has been utterly unrealized in any of the large compensatory education programs that have been evaluated so far."[25] The reason that Jensen provides for the failure of compensatory education programs is that the upper limit of one's intelligence is largely fixed by nature and hence not significantly alterable by educational intervention. Of course, to say that a human trait is largely fixed by nature is not to say exactly *how* that trait gets fixed. One could imagine that "nature" simply rolls the dice and randomly assigns a particular level of intelligence to each individual at birth. But as Herrnstein notes, this is emphatically not Jensen's view. Rather, Jensen forcefully argues that intelligence is very highly heritable; that is, *intelligence tends to run in families.* But notice: the assertion that intelligence is highly heritable is necessarily an assertion about *group differences* in intelligence. In the America of the early 1970s, the group difference in intelligence that most concerned everyone was the difference on IQ test scores of one standard deviation, or roughly fifteen points, separating the average black individual from the average white individual. Jensen's argument is that this *inherited* difference accounts for the achievement gap we see between "minority" and "majority" pupils. Thus Jensen managed to argue, probably more forcefully than any other academic at the time, that general intelligence is highly heritable, that its upper limit cannot be measurably increased through education, and that the difference between blacks and whites with respect to general intelligence and scholastic achievement is therefore due primarily to biological factors.

The key to Jensen's entire argument is the assertion that intelligence is highly heritable. Herrnstein spends considerable time in his article reviewing Jensen's evidence for this claim. I will not examine that review in detail, but because the truth or falsity of this claim is of such vital importance to society, I will discuss evidence for and against it in chapter four. For now, I want to concentrate on the question of why Herrnstein would want to include a discussion of such a controversial figure as Jensen in his own article.

To some extent, Herrnstein's discussion of Jensen's work was probably unavoidable. If you were writing an article in the early seventies on the current research into intelligence, and if you planned to have the article published in a national magazine with a fairly broad and educated audience, you would need to say *something* about the national figure who is most closely associated with intelligence research. That figure was Arthur Jensen. Although his *Harvard Educational Review* article may not have been widely read outside of academic circles, the controversy that it generated was real and widespread. As Herrnstein and Murray write in *The Bell Curve*, "During the first few years after the *Harvard Educational Review* article was published, Jensen could appear in no public forum in the United States without triggering something perilously close to a riot."[26] Although the controversy over Jensen's work may have been unavoidable and thus needed to be addressed in Herrnstein's own article, I suspect that the controversy was also useful to Herrnstein. Aside from making his own *Atlantic Monthly* article that much more controversial and hence widely read, the controversy surrounding Jensen's work allowed Herrnstein to make a larger point about the relationship between science and society. The point was that scientists must be allowed to address all questions freely and without fear of persecution. Although Herrnstein does not say so explicitly, the subtext of his article is that this view of science is especially important in a society that prides itself on being the heir to, and highest example of, Enlightenment thought.

But even if one grants this view of science, one may still want to argue that Jensen's work itself was not science but simply propaganda for the upper class. Herrnstein disagrees, of course, and sets out to defend Jensen's work. A large part of that defense consists in placing Jensen's work squarely within the context of mainstream scientific research on intelligence. Thus Herrnstein notes that Jensen's *Harvard Educational Review* article "is cautious and detailed, far from extreme in position or tone. Not only its facts but even most of its conclusions are familiar to experts."[27] Herrnstein goes on to note that "Jensen echoes most experts on the subject of IQ by concluding that substantially more can be ascribed to inheritance than environment."[28] The problem, as Herrnstein points out, is that many people, including perhaps many intellectuals, simply do not want to accept the results that science establishes. Thus, Herrnstein seems to be saying, what we may be witnessing with respect to the whole IQ debate, and particularly with respect to the reception of Jensen's work, is simply an expression of that all-too-common desire to kill the messenger because we do not like the message. As Herrnstein says, "Since the importance of inheritance seems to say something about racial differences in IQ that most well-disposed people do not want to hear, it has been argued that Jensen should not have written on the subject at all or that the *Harvard Educational Review* should not have, as it did, invited him to write on it."[29]

As that last quotation suggests, Herrnstein is clearly addressing multiple audiences. One of the most important of these is composed of academics and intellectuals like Herrnstein himself who populate humanities and social sciences departments in American universities and who decide what gets taught there and what gets published in scholarly journals. To this audience Herrnstein seems to be saying the following: If you believe what you are telling your students, if you believe the Enlightenment ideal that a society based on reason is always superior to one based on prejudice and superstition, then you had better be willing to go where reason leads you. To turn away from this path is to betray the academy and to evidence a certain failure of nerve.

And where do those whose nerve fails them turn? They turn of course back to prejudice and superstition. Nowhere is this more evident, Herrnstein seems to be saying, than in the scientific discussion of issues concerning innate racial differences, particularly in intelligence. We seem unable to confront this issue squarely and scientifically, in part because we have no faith that the ultimate answer we get will be the one we want. Hence we turn back to our popular prejudice in *equality*, understood not just in a political sense but in a biological sense as well. If science does not cooperate in this endeavor, we may even seek to subordinate science to politics, as Herrnstein clearly fears.

In a sense, then, Herrnstein's article represents a challenge to his colleagues. The challenge, which Gould and others accepted, is to enter into an open and scientific discussion of race and intelligence. In issuing this challenge, Herrnstein begins moderately, by conceding that, "although there are scraps of evidence for a genetic component in the black-white difference [in IQ scores], the overwhelming case is for believing that American blacks have been at an environmental disadvantage. To the extent that variations in the American social environment can promote or retard IQ, blacks have probably been held back." Then Herrnstein drops the other shoe:

> But a neutral commentator (a rarity these days) would have to say that the case is simply not settled, given our present stage of knowledge. To advance this knowledge would not be easy, but it could certainly be done with sufficient ingenuity and hard work. To anyone who is curious about the question and who feels competent to try to answer it, it is at least irritating to be told that the answer is either unknowable or better not known, and both enjoinders are often heard. And there is, of course, a still more fundamental issue at stake, which should concern even those who are neither curious about nor competent to study racial differences in IQ. It is whether inquiry shall (again) be shut off because someone thinks society is best left in ignorance.[30]

To some, I suspect, Herrnstein's apparent suggestion that the biology of race might play any part at all in group differences between IQ scores might be enough to indict his entire article and thus liberate the reader from the burden of finishing it. If so, this is clearly a rhetorical failure on Herrnstein's part. For the introduction and defense of Jensen's work sets up the third, and most important, argument Herrnstein makes. Herrnstein puts his argument in the form of a four-part syllogism that unfolds as follows.

1   If differences in mental abilities are inherited, and

2   If success requires those abilities, and

3   If earnings and prestige depend on success,

4   Then social standing (which reflects earnings and prestige) will be based to some extent on inherited differences among people.[31]

As Herrnstein notes, this syllogism has a number of critical implications for the future of America. The most important is that as American society progresses toward the salutary goal of guaranteeing equal opportunity to all, without regard to race, sex, religion, etc., society will tend to become more *un*equal than it has been in the past. This of course sounds completely counterintuitive. How could efforts to guarantee equality of opportunity *cause* inequality? Herrnstein provides an answer that seems, as one looks back over the last thirty years, remarkably prescient.

He begins by noting that increasing equality of opportunity will mean that those who in the past have been arbitrarily denied the opportunity to, for example, get an excellent education and thereby to better themselves will now be able to take advantage of those opportunities. This in turn will mean that society as a whole will benefit as the talents of more and more bright people that were underutilized in the past are now tapped. As the free-market conservatives like to say, the "economic pie" will increase as society draws on human capital that had been arbitrarily blocked from its most productive uses. The problem is that even as this pie increases, and even as more people have more resources in *absolute* terms than they had in the past, the *relative* difference in the amount of resources possessed by those at the top and bottom of the economic ladder will increase dramatically, owing to the fact that bright people who had been arbitrarily denied opportunity in the past will now be able to claim a larger share—their "fair" share—of the pie. The predictable result will be a society with more overall wealth and prosperity than it had in the past, but also a society with greater overall inequality. Such a society will also evidence higher levels of envy and correspondingly lower levels of fraternal feeling than it had in the past. To repeat, all of this comes about as the consequence of well-meaning attempts to increase equality of opportunity. Herrnstein drives this point home in the conclusion of his article when he writes, "Greater wealth, health, freedom, fairness, and educational opportunity are *not* going to give us the egalitarian society of our philosophical heritage. It will instead give us a society sharply graduated, with ever greater innate separation between the top and the bottom, and ever more uniformity within families as far as inherited abilities are concerned."[32]

Herrnstein's conclusion was particularly devastating to the contemporary liberal elites of the 1970s, for it was they who most wanted that which Herrnstein seemed to prove we can never have: an egalitarian society that guarantees equality of opportunity to all. Herrnstein's argument was that greater equality of opportunity in American society would in reality lead to the creation of a new caste system in America—one based not on birth or blood but on the equally arbitrary factor of inherited intelligence. Ironically, nowhere were the results of this new caste system more in evidence than in the elite northeastern American universities—the very places that, in post-sixties America, seemed to value egalitarianism above all. It turns out that during the late 1930s and throughout the 1940s most if not all of these elite universities, following the lead of educational reformers like James Bryant Conant (president of Harvard from 1933 to 1953), sought to transform their admissions processes from ones that explicitly favored the sons of wealthy northeastern families to ones that explicitly favored individuals who did well on standardized tests. To be sure, alumni preferences are still very much in force at America's elite universities, but as Nicholas Lemann points out in *The Big Test: The Secret History of the American Meritocracy*, reformers like Conant were remarkably successful at literally changing the face of American higher education. In particular, Conant and a handful of others were instrumental in institutionalizing the use of standardized tests as a significant factor (perhaps, ultimately, the most significant factor) in determining admissions to American colleges.[33] Thus, Lemann argues, reformers like Conant were the ones most responsible for establishing an American meritocracy.

They did not call it that, of course, for the term *meritocracy* was not coined until 1958. It was then that the "chief thinker" (to use Lemann's characterization) of the British Labour Party—a man by the name of Michael Young—published a book entitled *The Rise of the Meritocracy*. Young used the term to refer to rule by an intellectual elite. He saw this rule as all but inevitable after the passage in England of the Education Act—a law that opened up educational opportunity to the lower class and that was championed by the Labour Party. But Young's book was not an attempt to praise the coming meritocracy in England. Rather, it was very much an attempt to bury the concept. For, as Lemann notes, Young saw much more clearly than his Labour Party colleagues the "inherent contradictions" between a society based on the concept of equality of opportunity and one based on the old-fashioned Labour Party ideal of egalitarianism. Young understood that a meritocracy would only replace an old elite with a new elite. This new elite would, however, be much more firmly entrenched in power than any of its predecessors throughout history because its members would possess something (intelligence) that really made a difference. Young also understood that a meritocracy would isolate and stigmatize the lower class. Finally, he understood very well that the worst effect to come from the rise of the meritocracy would be the fatal undermining of the *rhetoric* of the Labour Party. As Lemann explains, Young saw that "with equal opportunity for all, inequality would increase and the Labour Party would be powerless to do anything about it because its central argument, that British society was constructed unfairly, had been taken away."[34]

Of course, this is not *quite* right. One can still argue against a meritocracy on the grounds that it is unfair. One simply needs to point out that from a moral point of view a given individual no more deserves to be born with a high natural intelligence than he or she deserves to be born into a wealthy family. The logical extension of this argument is that from a moral point of view no one deserves *anything* that he or she possesses, either in the form of innate qualities or material goods. Indeed, even if all the wealth you have has come to you *solely* through your own hard work, you still do not deserve it, because you do not deserve to have been born with the character traits that enabled (or, more accurately, *caused*) you to work hard.

If that sounds a bit odd, it is worth reflecting on this: The argument I have just sketched forms the very foundation of *A Theory of Justice*, a book written by the late John Rawls, a Harvard philosopher, and published in the same year as Herrnstein's *Atlantic Monthly* article. *A Theory of Justice* is arguably the most important work of political philosophy written in English in the twentieth century. It is a weighty philosophical tome that serves as the principal intellectual justification for a particular social and economic order. One could do worse than say that *A Theory of Justice* is to modern welfare-state liberalism what Adam Smith's *Wealth of Nations* was to laissez-faire capitalism and what Karl Marx's *Das Kapital* was to communism. Rawls argues that, because no one deserves any of the natural endowments (like intelligence or temperament) with which he or she was born, and because these endowments have not been distributed "fairly" by nature, the most just society is one which treats these natural endowments as belonging to the society as a whole and not the individual. Rawls grants the obvious point that these natural endowments cannot *themselves* be redistributed by society, but he argues that the differential effects that these endowments produce can be adjusted by societal arrangements so that these effects equal out for everyone in society.

What would such a society look like? Again Rawls grants the obvious point that no society can be *economically optimal* unless no person can be made better off without making at least one person worse off.[35] But there may be many such economically optimal societies with many different economic arrangements. Justice demands, according to Rawls, that we select from among these societies the one that provides the greatest benefit to the least advantaged member of the society. If this criterion is met and there are two or more societies that provide equally great benefits to their least advantaged member, we select the one that provides the greatest benefit to the next least advantaged member, and so forth. This is, roughly stated, Rawls's *difference principle*.[36] It is also, essentially, a description of the perfectly administered welfare state. And, finally, the vision Rawls develops of the just society is *strikingly* compatible with the old aristocratic sense of noblesse oblige. Members of the new elite are allowed to keep their natural privileges—now simply called *assets*—but they are made to use these assets in the service of the less fortunate.

As I have said, *A Theory of Justice* functions as something of a philosophical *apology* for modern welfare-state liberalism. But there is a rhetorical problem with Rawls's book. (Actually, there are several. But one is especially relevant to us.) Rawls does not provide a convincing argument for one of the central claims in his book—the claim that no one deserves his or her natural endowments.[37] Now, over the course of the last twenty years or so I have become convinced that to be a contemporary liberal it is necessary, although not sufficient, that one believe just this claim. I have also become convinced that in a theoretical sense this claim must be true. But try convincing a group of undergraduates of this, and you begin to see the problem that contemporary liberalism faces. Even individuals who might benefit from a society in which everyone believed this cannot themselves seem to embrace the belief.

Let me put the rhetorical problem here more concretely. It is, I would argue, one thing to say, "You were born rich and inherited a large trust fund, while I was born poor and inherited nothing; thus I should be entitled to some of your money because you really do not deserve it." It is quite a bit more problematic to say, "You were born smart and therefore able to make a lot of money, while I was born dull and therefore not able to make a lot of money; thus I should be entitled to some of your money because you really do not deserve it." And, finally, it is downright bizarre to say, "You were born with a temperament that made you disciplined and therefore enabled you to work hard and make a lot of money, while I was born with a temperament that made me lazy and therefore not able to work hard and make a lot of money; thus I should be entitled to some of your money because you really do not deserve it." From

Rawls's perspective all three of these claims are equally valid. But they surely are not equally persuasive.

This brings us back to Young's insight about the rhetorical problem faced by the British Labour Party. Young correctly saw that while it is relatively easy to argue for a meritocracy on the basis that no one really deserves to inherit wealth, it is much more difficult to argue for measures to improve the lot of the poor once a meritocracy is fully in place. Somehow one needs to find a way to convince the intellectual elite that they need to do something, beyond simply guaranteeing equality of opportunity, to help those who cannot compete in a fair system. In other words, one needs to find a way to maintain a sense of community and fraternal feeling in a meritocracy, even as the workings of the meritocracy tend naturally to drive the class composed of cognitive elites farther and farther from the class composed of their less intellectually endowed fellows. The problem is compounded and perhaps made intractable by the fact that in a meritocracy it is all too easy for the intellectual elite to think that the fault is no longer with the system but rather with nature, which cannot be changed and which has just decided to allocate natural endowments unfairly. In other words, Mother Nature is not democratic.

In a sense, Herrnstein's *Atlantic Monthly* article reprises and updates Michael Young's argument in *The Rise of the Meritocracy*. In so doing, Herrnstein places contemporary liberals in something of a double-bind. On the one hand, contemporary liberals cannot really argue *against* equality of opportunity (as Young thought the Labour Party should have done in the 1950s), for that concept is just too firmly engrained in the American ethos. Additionally, to argue against equality of opportunity—and to embrace the Rawlsian notion that no one deserves what he or she has—would put contemporary liberals in the uncomfortable position of needing to concede that (for example) disadvantaged minority individuals who managed to make it into elite universities did not really deserve to be there because, again, *no one did*. But this just sounds silly, and perhaps a bit insulting.

On the other hand, contemporary liberals could not really embrace the concept of equality of opportunity because that would seem to put them in the uncomfortable position of needing to concede that white individuals really did deserve to be at elite universities because they got there by coming out ahead in a fair competition. Of course, contemporary liberals might embrace equality of opportunity in theory but argue that it had not yet been achieved in practice. They could say, as they do, that the playing field is not yet level, but that a level playing field is what we should all be striving to achieve. The problem with saying this, however, is that even in theory there is an inherent contradiction between equality of opportunity and egalitarianism. More specifically, Rawls's difference principle is flatly incompatible with the principle of meritocracy because the difference principle *requires* that we *not* assign rewards and benefits to individuals based on their natural talents, but rather, that we assign rewards and benefits to individuals based on an analysis of what would be best for the least advantaged member of society. In his *Atlantic Monthly* article Herrnstein simply pointed out this incompatibility, using IQ as an example of a natural talent.

Of course, Herrnstein was certainly not the first to note the tension between meritocracy and egalitarianism. Nor was he even the first to do so from within the walls of Harvard Yard. Eight years before Herrnstein published his *Atlantic Monthly* article, Clark Kerr delivered the Godkin lectures at Harvard. When Kerr delivered these lectures, later published as *The Uses of the University,* he was in his fifth year as chancellor of the University of California system.[38] As his audience surely knew, during those five years Kerr had managed to make the University of California into the preeminent public university system in America, and to make its flagship campus, Berkeley, into a rival even of Harvard itself. Also, as his audience surely knew and appreciated, Kerr managed to do all of this by hewing closely to the educational philosophy set down two decades earlier by James Bryant Conant. In particular, Kerr recognized the value of elitism in a university. But he also recognized the danger of recognizing this value in American society. He put this paradox to his Harvard audience in the following way: "The great university is of necessity elitist—the elite of merit—but it operates in an environment dedicated to an egalitarian philosophy. How may the contribution of the elite be made clear to the egalitarians, and how may an aristocracy of intellect justify itself to all men?"[39]

The questions that Kerr poses are both vital and *perfectly* formulated. In his *Atlantic Monthly* article Herrnstein offered at least a tentative answer to the second question. He noted that an aristocracy of intellect might be justified on the grounds that there is now good scientific evidence to establish that there are significant biological differences between individuals, particularly with respect to intelligence. The egalitarians at Harvard would have none of it. They insisted that biological differences should have no place at Harvard, which was fine, because this insistence acknowledged a certain equality among scholars. But they also insisted that biological differences should have no place outside of Harvard Yard, which was not fine, because that insistence made it difficult to explain why *everyone* should not then be allowed into Harvard.[40]

Of course, Richard Herrnstein was not the only controversial professor at Harvard during the 1970s. A scant two years after Herrnstein published *IQ in the Meritocracy*—the book-length version of his *Atlantic Monthly* article—another Harvard professor began to stir up controversy in Cambridge and beyond.[41] Let us now examine his story.

CHAPTER THREE

Edward O. Wilson Brings More
Controversy to the Yard

What Shakespeare said of mortals may be true of books as well. "Some are born great, some achieve greatness, and some have greatness thrust upon 'em." When in 1975 the Belknap Press of Harvard University Press published *Sociobiology: The New Synthesis*, the book's author, Edward O. Wilson, a Harvard zoology professor, immediately found himself at the center of a controversy that was every bit as heated as the one that had engulfed Herrnstein. To some, *Sociobiology* was destined to be great. It was the new synthesis of the natural and social sciences for which everyone had been waiting. To others, *Sociobiology*'s central premise at first appeared dubious, but then achieved acceptance, and perhaps even enthusiastic embrace, because it really could explain so much so well. And, finally, to still others *Sociobiology*'s greatness was simply the result of a shameless media and public relations campaign on the part of Harvard University Press.

With that said, *Sociobiology* is a difficult book to describe. It is first and foremost a large book, both in the number of its pages (nearly seven hundred) and in its trim size (an odd twelve inches square). The "twenty-fifth anniversary edition," which I have before me, sports on its front cover a full-color photograph of two elephants playfully cavorting on what may be (for all the reader knows) the African savanna. The photograph is the entire cover, over which is placed text and three smaller color photographic inserts of penguins, chimpanzees, and weevils. Inside one finds almost a dozen original drawings— "compositions" in black and white, as they are described—which span facing pages, thus forming large twelve-by-twenty-four-inch canvases. These compositions depict imagined scenes of specific animal societies in the wild, "with as many social interactions displayed as can plausibly be included in one scene."[1]

The photographs and original drawings give *Sociobiology* something of the flavor of a coffee-table book. It is a conversation piece—something every intellectual should own. But at the same time, the book contains numerous charts and graphs, and a fair number of mathematical equations. Add to this the fact that the text is presented in two-column format, and *Sociobiology* begins to look very much like a biology textbook. It becomes daunting, but remains interesting throughout. In the end, the best description I can think of for *Sociobiology* is one that I readily admit is much more impressionistic than analytical. I would say that *Sociobiology* is quintessentially the type of book that people read "in" rather than read.

But what made this book and its author so controversial? The answer can be found in the book's first and last chapters. On the second page of the first chapter Wilson offers a definition of the discipline he seeks to bring into existence: "*Sociobiology* is defined as the systematic study of the biological basis of all social behavior."[2] Three years later, in *On Human Nature*, a book that can properly be called the sequel to *Sociobiology*, Wilson repeats his earlier definition almost exactly, but with a minor addition. In the introductory part of *On Human Nature* Wilson tells us of "the new discipline of sociobiology, defined as the systematic study of the biological basis of all forms of social behavior, in all kinds of organisms, including man."[3] There—Wilson had said it. But back in 1975 no one had missed the point. Sociobiology was not just, or even primarily, about the lower animals. Either it was principally about humankind or it was just not worth getting upset about.

Those who chose to criticize sociobiology on political grounds—and they were legion, even as early as 1975—interpreted Wilson's new synthesis in an entirely straightforward manner. They saw in sociobiology simply the latest attempt by science to reduce human beings to mere biological entities. In so doing, they argued, sociobiology must necessarily rob humankind of its soul. That was bad enough, of course. But, historically, scientific explanations of human nature always tended to favor conservative political ideologies. Such explanations always tended to be on the wrong side of debates about whether women should be granted the same rights as men, or about the workability of socialist economic systems, or even about the possibility of a lasting peace among nations. In short, history offered Wilson's political critics ample reasons to think his new synthesis should be seen as simply the same old story.

I have been over this ground in chapter one, where I examined some of Gould's various examples of "bad" science throughout history and at the same time critiqued his own attempt to historicize science. It is not my intention to go over these arguments again. But I do need to point out that for those who believe science has no business intruding into the domains of morality and ethics, *Sociobiology* is a book that cannot be ignored—especially because of its concluding chapter. There are at least two ways to get to this chapter. One can plough through several hundred pages of interesting but often excruciatingly dense analysis of everything from polygenes and linkage disequilibrium to temporary social parasitism in insects to the ecology of social behavior in primates. Or one can simply flip to the back of the book. Either way, one arrives at a chapter provocatively entitled "Man: From Sociobiology to Sociology." Here Wilson seems to take reductionism to new heights. In a section on human ethics, he immodestly suggests that "scientists and humanists should consider together the possibility that the time has come for ethics to be removed temporarily from the hands of the philosophers and biologized."[4] Wilson does concede that science does not yet know enough about the human brain to be able to design a culture that could eliminate unhappiness. That, he says, will have to wait for significant advances in neuroscience. Wilson also concedes that "a genetically accurate and hence completely fair code of ethics must also wait."[5]

But the assumption in the second concession (which hardly seems like much of a concession) is that with a genetically accurate understanding of human beings a completely fair code of ethics will become possible, if not self-evident. When we ask what the knowledge of genetics can tell us about the relative fairness of ethical systems, the answers that emerge from Wilson's last chapter seem to confirm the worst fears of those who say that when science ventures into the domains of morality and ethics the results are often disastrous for liberal and progressive politics. Thus, at various places throughout the last chapter of *Sociobiology*, Wilson maintains all of the following theses: that human tendencies toward "aggressive dominance systems, with males generally dominate over females," as well as tendencies for "intensive and prolonged maternal care" represent difficult to alter biological traits that have persisted since the days of our primate ancestors; that the nuclear family is "the building block of nearly all human societies," with other arrangements, such as those found in Israeli kibbutzim, being evolutionarily unstable; and that our biological tendencies toward territoriality and tribalism represent a "problem," insofar as these tendencies are poorly adapted "for the extended extraterritorial relationships that civilization has thrust upon" us, with the result sometimes being xenophobia and war.[6]

What I have just sketched is, I would argue, the best case one can make for the politically conservative nature of *Sociobiology*. For balance, I should add that there is much in the book, and in particular in its last chapter, that seems not to support any particular political agenda and that may serve to undermine some conservative political arguments. Indeed, Wilson has practically nothing to say in *Sociobiology* about the hottest of the hot-button issues: human intelligence and, in particular, its connection to race. The only real mention Wilson makes of human intelligence understood as an individual, measurable quality comes in the last chapter, where the whole discussion receives only about a dozen lines. To be sure, Wilson does reference Herrnstein's *Atlantic Monthly* article "I.Q.," and he does note its central thesis: that equalizing environmental opportunities in a society may mean that "socioeconomic groups will be defined increasingly by genetically based differences in intelligence."[7] Yet Wilson immediately qualifies this point by adding that "despite the plausibility of the general argument" for the stratification of societies based on inherited traits, there is little evidence that this has actually happened in human history. With respect to human intelligence specifically, Wilson notes, "The hereditary factors of human success are strongly polygenic and form a long list, only a few of which have been measured. IQ constitutes only one subset of the components of intelligence. Less tangible but equally important qualities are creativity, entrepreneurship, drive, and mental stamina."[8]

This language is a far cry from the claims Herrnstein and Jensen were making in the early 1970s. Wilson even references the work of Richard Lewontin on human diversity, perhaps as a way of inoculating himself against the charge of racism. Lewontin, whom we will meet again very shortly, was at the time of *Sociobiology*'s publication a well-established population geneticist at Harvard. In a 1972 article entitled "The Apportionment of Human Diversity," he reports on an extensive statistical analysis he did on genetic factors associated with blood samples from several racial and ethnic populations. As Wilson notes, Lewontin "found that 85 percent of the variance [within these samples] was composed of diversity within populations and only 15 percent was due to diversity between populations."[9] Lewontin's blood factor analysis of genetic difference became central to the debate about race and biology that occurred throughout the 1970s. Although there are many ways of interpreting Lewontin's findings, one of the most common highlights this statistical fact: there is vastly more genetic variation between two randomly selected individuals in any given racial or ethnic population than between the statistical averages of the genetic variation of all individuals of one racial or ethnic population and the genetic variation of all individuals of a different racial or ethnic population. That is probably the most accurate way of putting the relevant point. Another way of saying much the same thing is to note that there is much more difference between any two randomly selected white individuals than between two statistically "constructed" individuals, one white and one black, each of whom represents the *average* of his race. One approach to take in light of this fact is to say that the concept of race simply has no useful role to play in the science of human evolutionary biology. Wilson was clearly on board with this approach. He emphasizes this very point in the last chapter of *Sociobiology*.[10]

But none of that seemed to matter. Almost immediately after the publication of *Sociobiology*, a group of individuals from the Boston and Cambridge area formed something called the Sociobiology Study Group. The aim of the group was to facilitate the development and dissemination of scientific arguments against sociobiology that would appeal to an audience of educated individuals not specifically trained in the natural sciences. The members were principally scientists and doctors, mostly associated with Harvard or MIT, including Stephen Jay Gould and Richard Lewontin. Ironically, Lewontin (also a Marxist) was brought to Harvard in the early seventies principally at the request of Wilson himself. The connection between the two men—both of whom were born in 1929—went back to the sixties, when they were part of a small group of Young Turks who were hoping to move the discipline of biology more completely toward a synthesis of population genetics and evolutionary ecology, precisely the kind of synthesis Wilson presents in *Sociobiology*. Lewontin was one of the preeminent mathematicians in the group. One of his specialties was game theory, which he was supposed to use to demonstrate how the findings of evolutionary ecologists like Wilson could stand up under mathematical scrutiny. Thus one way of interpreting Wilson's motive in bringing Lewontin to Harvard is to see it as a straightforward attempt to consolidate his forces. According to Segerstråle, in *Defenders of the Truth*, Wilson apparently had no inkling that Lewontin would turn against him. This may well be plausible, for as Segerstråle also notes, there is little doubt that Lewontin became affiliated with the Sociobiology Study Group in large part because of the IQ controversy swirling around Herrnstein at about the time Lewontin came to Harvard. Herrnstein's theories on IQ would of course make a socialist-egalitarian world less likely. Lewontin clearly saw this and apparently resolved to stamp out biological determinism wherever it was to be found. In Lewontin's view, sociobiology happened to be a conspicuous example of such biological determinism.[11]

The problem with interpreting the relationship between Lewontin and Wilson in this way, however, is that it makes Wilson look awfully naïve. Wilson, who clearly knew Lewontin was a Marxist, could not have been unaware that it might be difficult for him to separate his science from his politics, because Marxism generally requires more of a political commitment than other ideologies. Also, Wilson must have understood at some level that sociobiology, although unrelentingly materialist, was simply not congenial to the utopian idealism espoused by the Marxists. Thus one might conjecture that, somewhere deep in his unconscious, Wilson's motive in bringing Lewontin to Harvard may have owed *something* to Nietzsche's maxim that one should always try to cultivate one's enemies. The classically Nietzschean point here is that only by overcoming worthy adversaries can one demonstrate his or her own superiority.[12] Fortunately for Wilson, the study group did include some very worthy adversaries. In addition to formidable intellectual figures like Gould and Lewontin, the membership list also included Ruth Hubbard, a Harvard professor of biology, and Stephen Chorover, a professor of psychology at MIT.

But in addition to these established names in science, the Sociobiology Study Group also found room for those without fame or without even recognizably obvious scientific credentials. For example, the group included a self-described "pre-med" student at Brandeis University and a teacher at a local public high school. The inclusion of these obviously less scientifically noteworthy individuals may have helped to connect the group more closely to the "common" people. It may also have served the rhetorical function of bolstering their argument that scientific expertise alone should not be allowed to settle public policy questions.

Finally, I should note that there were some very conspicuous *nonmembers* of the Sociobiology Study Group. Perhaps the most important of these was Noam Chomsky, a professor at MIT, another Marxist, and the one individual who might well be credited with having invented the modern discipline of linguistics. Chomsky was apparently invited to be a member, but declined. The reason for Chomsky's refusal to join is illuminating. As Segerstråle notes, while most of the members (including especially the Marxists) believed the standard social science creed that a meaningful biological human nature did not exist, and that what some took to be human nature was really no more than the soft wax upon which the various forces of language and culture impressed themselves, Chomsky had different ideas. He apparently wanted to recover the early Marx—the Marx of the *1844 Manuscripts*—who believed that a more or less fixed human nature, expressed as our *species being*, really did exist, and that only with an understanding of our species being could we fashion a truly humane society. The point of this version of Marxism was precisely that human nature is *not* infinitely malleable. Chomsky believed that abandoning the very idea that a more or less stable human nature existed and could be known would do more harm than good to the suffering masses because it would undercut the argument that humans had real needs and desires that the oppressors might be violating—and replace it with the idea that human suffering might just be all in the mind. On the other hand, the study group members apparently thought that embracing the idea that a more or less stable and knowable human nature existed was just too dangerous because such an embrace would most likely serve the interests of the white Western male patriarchal power structure.[13] That, at any rate, was the argument that the Sociobiology Study Group made in its bitter battle with the author of *Sociobiology*.

The study group drew first blood in this battle with the publication of an open letter that appeared in the 13 November 1975 edition of the *New York Review of Books*.[14] The letter, signed by sixteen members, begins by asserting that Wilson's book "has little relevance to human behavior" and that its "supposedly objective, scientific approach in reality conceals political assumptions."[15] The authors go on to detail a series of supposedly "scientific" objections to Wilson's theories. Some of these objections are plainly misrepresentations of Wilson's actual theory. For example, in a discussion of the relationship between evolution and social behavior, it is alleged that "for Wilson, what exists is adaptive, what is adaptive is good, therefore what exists is good."[16] Thus the authors can accuse Wilson of thinking *and claiming* that the present society is "good," even though women, for example, are underrepresented in positions of "power" within society.

But Wilson does not claim that this is in any sense "good." He and the evolutionary psychologists who built on his work claim *only* that the human traits that exist today—traits like sexual jealousy or a taste for fatty foods—

were adaptive and, yes, in some narrow survival sense "good" *in the ancestral environment*. But such traits could very well be *maladaptive* today, as a glance at heart attack rates in industrialized countries immediately makes clear. Whether traits like sexual jealousy are adaptive or maladaptive today is certainly open to *political* discussion. But what is not open to political discussion, according to Wilson, is that such traits exist today because they were once, long ago, adaptive.

The study group authors are on firmer ground in their criticism of Wilson when they resort to the traditional attacks that have been made against biologists and other natural scientists who venture into the domain of the social sciences. Hence, at various points in their letter, they insist that one cannot draw any meaningful inferences about human behavior from animal studies. This is an important and relevant point, given that almost all of *Sociobiology* is devoted to an examination of nonhuman animal behavior. It is of course important to point out the ways in which "slavery" in ant colonies is obviously different from slavery that was practiced and condoned by human societies. Picking up on this point, the study group authors claim, in opposition to Wilson, that each and every human culture is not only untethered from biological constraints, but also unlike any other human culture. Thus, in a concession that is not much of a concession after all, the authors write: "We suspect that human biological universals are to be discovered more in the generalities of eating, excreting and sleeping than in such specific and highly viable habits as warfare, sexual exploitation of women and the use of money as a medium of exchange."[17]

If there is a politics lurking behind Wilson's book, there is just as surely a politics behind the criticisms offered by the sixteen members of the Sociobiology Study Group. It is exactly the kind of politics one would expect from a group dominated by post-sixties leftist intellectuals. The study group authors simply will not concede that there are any significant biological limits to the creation of any human culture, because conceding any such limitation would necessarily lessen the power of the intellectuals to bring about change through politics based on the particular kind of social engineering that appealed so much to them.

To be sure, the study group authors do attempt at various places in their letter to explain what they see as the misapplication of science to what are essentially political questions. But in their zeal to keep science, and particularly biology, out of politics the authors grossly exaggerate the dangers of sociobiology. Indeed, try as they might, they simply cannot resist invoking the specter of Hitler and Nazism. Thus a mere two paragraphs into their letter, scientific criticism gives way to political propaganda as the authors argue that the biological determinism allegedly at the heart of sociobiology "provided an important basis . . . for the eugenics policies which led to the establishment of gas chambers in Nazi Germany."[18] They then close their letter with a classic example of the ad hominem fallacy by insinuating that Wilson's theories are suspect precisely because they cannot easily be separated from "the personal and social class prejudices of the researcher."[19]

For his part, Wilson was not about to take the criticisms of his magnum opus lying down. Less than a month after the study group's letter was published Wilson shot back with a reply, also printed in the *New York Review of Books*, which accused his critics of distorting his theory and misinterpreting crucial quotations from his book. Although more measured in its tone than the attacks made by his critics, Wilson's reply displayed a feistiness all its own. At the conclusion, Wilson laments the fact that, although his office is located in the same building as those of several members of the Sociobiology Study Group, not one individual from that group ever came to him to discuss any aspect of sociobiology. Warming to his point, Wilson then accuses the members of engaging in a kind of "self-righteous vigilantism," which, by grossly misrepresenting the hard work of scientists like himself (the Nazi reference is clearly on his mind here), only serves to chill "the spirit of free inquiry and discussion crucial to the health of the intellectual community."[20]

Despite the bitterness evidenced by these two letters, the exchange between the Sociobiology Study Group and Wilson did involve some real issues. Thus this is the appropriate time to repeat a point I have made several times. Scientists are of course human and subject to human weakness. Some scientists, like Josef Mengele, are also *evil*, in the most basic sense of the term. Certainly nothing I have said should be interpreted as meaning that scientists as a group are less likely than the rest of us to do bad things. Hence the Sociobiology Study Group was correct to point out that science has often been used to justify plainly unjust practices. They were also right to warn us to be on the lookout for faulty justifications in our own time. But Wilson was also correct to note that unwarranted personal attacks on scientists serve only to diminish scientific inquiry and the search for truth. Such attacks, as Wilson clearly implied, have a chilling effect on all scientists.

Wilson was to find this out the hard way in a famous incident that took place in February 1978 at a special academic convention sponsored by the American Association for the Advancement of Science. The incident in question has now entered into the folklore of sociobiology, if not science itself. Although each of the four accounts I have come across puts a slightly different spin on what happened, the broad facts seem not to be in dispute. The convention, which was held at a Washington, D.C., hotel, was called for the explicit purpose of bringing together advocates and critics of sociobiology so that all parties could engage in a frank scientific discussion of the issues raised by Wilson's new synthesis. Like other academic conventions of its sort, this one included a number of sessions at which speakers presented various papers and then took comments and questions from the audience. One session featured as speakers both Stephen Jay Gould and Edward O. Wilson. This session was held in the ballroom of the hotel and was apparently very well attended. The incident occurred here. One of the most detailed descriptions of the incident is provided by Segerstråle, who was in the audience at the time:

> The session has already featured Gould, among others, and Wilson is one of the later speakers. Just as Wilson is about to begin, about ten people rush up on the speaker podium shouting various epithets and chanting: "Racist Wilson you can't hide, we charge you with genocide!" While some take over the microphone and denounce sociobiology, a couple of them rush up behind Wilson (who is sitting in his place) and pour a jug of ice-water over his head, shouting "Wilson, you are all wet!" Then they quickly disappear again. Great commotion ensues but things calm down when the session organizer steps up to the microphone and apologizes to Wilson for the incident. The audience gives Wilson a standing ovation. Now Gould steps up to the microphone saying that this kind of activism is not the right way to fight sociobiology—here he has a Lenin quote handy, on "radicalism, an infantile disorder of socialism." For his valiant handling of the situation, Gould, too, gets a standing ovation. (The audience does not quite know how to react to any of this but applauding seems somehow right.) Wilson—still wet—gives his talk, in spite of the shock of the physical attack. He explains his own non-political background motivation for *Sociobiology* and produces various types of studies in support of the idea of a genetic basis for human behavior. After all the action, his calmly delivered talk is something of an anticlimax.[21]

Most accounts I have read, including Segerstråle's, identify some of the disrupters as member of the International Committee Against Racism. Other disrupters may have belonged to a group called the Sociobiology Study Group of Science for the People, which was formed out of the merger of the original members of the Sociobiology Study Group and the members of the Boston chapter of Science for the People. This latter group—Science for the People—has been described by Segerstråle as "a national forum for left-wing academic activism."[22] If nothing else, the merger created a new group with an impressively long name and a national base. But regardless of who Wilson's attackers were, or what group or groups they ultimately belonged to, the incident does seem to dramatize the personal nature of the sociobiology controversy. In truth, it may also have helped to generate some sympathy for Wilson, if not for his scientific theory. It almost certainly discredited Wilson's critics, at least among most members of the general public.

While this attack on Wilson was absolutely unjustified, under any circumstances, the obvious question remains: What, if anything, was the connection between Wilson's politics and his science? This is certainly a fair question in the present context, for at least two reasons. First, this same question has already been raised with regard to other scientists like Gould and Lewontin. Second, Wilson, like Gould, but unlike Lewontin, is prone to remarkably high levels of self-disclosure in his work. He seems to want us to understand where he is coming from, almost on a personal level. Perhaps to further this understanding Wilson took the unusual step (for a natural scientist) of writing an autobiography, which he titled *Naturalist*, and which was published in 1994.[23] Of course, one should be careful not to believe everything one reads in an autobiography. The genre is sometimes fairly close to fiction. But even without examining too closely what Wilson says in *Naturalist*, several contours of his politics and personality emerge quite clearly, both from his other writings and from his colleagues and critics.

Politically, Wilson, though no Marxist, was what one might call a standard Harvard liberal, at least during the 1970s. In fact, Tom Wolfe describes Wilson as "a conventional liberal, PC, as the saying goes."[24] Wolfe, it should be noted, is clearly one of Wilson's biggest fans. Ullica Segerstråle (another fan) notes that throughout the sociobiology controversy of the 1970s "Wilson prided himself on being a fairly liberal thinker—the sort of person who naturally falls to the left of center politically."[25]

But what about Wilson's own writings? To be sure, at the conclusion of *Sociobiology* one does *not* find a call for the workers of the world to unite. Instead, one finds a kind of pragmatic resignation to "the planned society—the creation of which seems inevitable in the coming century."[26] Reading *well* between the lines here, it seems that, in the mid-seventies at least, the only real political commitment Wilson could manage was a halfhearted embrace of the "end of ideology" thesis made popular a decade and a half earlier by Harvard's Daniel Bell and instantiated in the early seventies by John Rawls, in A Theory of Justice. One of the best examples of the type of political philosophy that flowed from the "end of ideology" thesis was offered by a Harvard graduate, John Kennedy, in a 1962 commencement address he delivered, ironically, at Yale. In speaking of the challenges—particularly the economic challenges— that lay ahead, Kennedy said: "What is at stake in our economic decisions today is not some grand warfare of rival ideologies which will sweep the country with passion but the practical management of a modern economy."[27] In other words, the future belonged not to those who thought in terms of grand class struggles leading to utopian revolutions, but to those who thought in eminently pragmatic terms about effective solutions to intricate economic and social problems. The point was that at the end of ideology we would need good managers, not inspiring revolutionaries.

Needless to say, this vision of politics did nothing for Marxists like Gould and Lewontin. Nor did it excite post-sixties radicals on the cultural left. It was all too mature and sensible. But it was probably the best description of Wilson's politics, to the extent that any type of politics could be discerned from Sociobiology. I would argue, however, that finding a politics in Sociobiology was not as easy as many of Wilson's critics made it seem. Even though the last chapter of that book does deal with humans, the treatment is still too sketchy to discern with any precision Wilson's likely position on public policy questions.

To some extent that situation changed in 1978 with the publication of On Human Nature, a book that, as I have said, functions as the sequel to Sociobiology. On Human Nature cannot help but provide a fuller understanding of Wilson's politics, which is still left of center but thoroughly pragmatic. Even so, this apparently did not satisfy Wilson's liberal critics. Consider Wilson's position on the issue of male-female sex roles in society and his critics' response to that position. This issue had come to the fore in the late seventies as the Equal Rights Amendment was being debated in states across America and as women were moving into domains of society that had once been occupied exclusively by men. On Human Nature addresses this issue at some length. Wilson begins by noting that it is probably in our power to fashion a society that would effectively eliminate all or nearly all differences in behavior between the sexes. Thus we could probably create a society in which equal percentages of males and females cared for very young children, and equal percentages of males and females ran major corporations. But Wilson argues that to bring this society into being "the amount of regulation required would certainly place some personal freedoms in jeopardy, and at least a few individuals would not be allowed to reach their full potential."[28] Of course, as Wilson also notes, if we decide *not* to decide to attempt to eliminate sex-role differences in society—if we decide simply to guarantee equality of opportunity for all and then let the chips fall where they may—the result would almost certainly be a society in which "men are likely to maintain disproportionate representation in political life, business, and science. Many [men] would [also] fail to participate fully in the equally important, formative aspects of child rearing."[29] Thus we face a dilemma brought about by our desire to embrace simultaneously the principles of egalitarianism and the principles embodied in a system which merely guarantees equality of opportunity. This dilemma is, obviously, very similar to the one Herrnstein identifies regarding egalitarianism and meritocracy.

But, as Segerstråle notes, the very fact that Wilson named this dilemma as such was largely what got him into such trouble with liberals and feminists during the seventies.[30] At that time liberals and feminists were insistent that we did not have to make a choice between egalitarianism and simple equality of opportunity understood as strictly fair treatment of men and women. It was precisely this insistence that, in the early seventies, caused the doors of our military academies be open to all, regardless of sex. Egalitarianism demanded it. But when the average female cadet could not meet the same physical training requirements as the average male cadet, it was again precisely this insistence that we not be forced to choose between egalitarianism and equal treatment that compelled the lowering of physical training requirements for all and (when that still did not quite work) the institution of gender norming (a practice that adjusts the scores on physical training tests so that men and women *appear* to be equally capable of doing the kinds of things we expect soldiers to do).[31]

The point that Wilson was making in regard to sex-role differences—a point that seemed wholly lost on his critics—was simply that by understanding the *slight* (his word) but biologically innate differences between males and females we could better delineate "the options that future societies may consciously select."[32] This was a sensible and pragmatic point. The problem was that in the seventies it was not liberal enough.

Times change, however. And to some extent, so do individuals' politics. It is worth noting here that the trajectory of Wilson's own politics can reasonably be seen to follow that of many Kennedy liberals who, in the three decades since the end of the sixties, slowly turned themselves into *neo-conservatives*. Thus if we examine Wilson's 1998 book *Consilence*, we find that its last chapter simply has a different feel to it than the last chapter of *Sociobiology*. There is, thankfully, no more talk of "the planned society." Its "inevitability" apparently collapsed with the fall of communism and the unabashed victory of liberal democracy. Instead, Wilson is now more *explicitly* concerned with conservation, both of our human traditions and of our planet's genetic diversity. In the final pages of *Consilence* Wilson predicts, optimistically, that as the new millennium begins humankind will "come to understand the true meaning of conservatism. By that overworked and confusing term I do not mean that pietistic and selfish libertarianism into which much of the American conservative movement has lately descended. I mean instead the ethic that cherishes and sustains the resources and proven best institutions of a community. In other words, true conservatism, an idea that can be applied to human nature as well as to social institutions."[33] This sounds very much like the neo-conservatism that is ascendant today. In that sense, Wilson is politically very much in step with the times.

But enough of Wilson's politics. What has been said thus far certainly tells us less about Wilson the man—his personality and what makes him tick—than a similar examination of political concerns would tell us about Gould or Lewontin. For, unlike his Marxist colleagues, Wilson's political views, though important, seem plainly secondary to what are probably the two defining aspects of his personality: his love of nature and his struggle (let us call it that) with his Christian upbringing. (One might also mention, as a third aspect, his substantial ambition. But I think that needs fairly little explanation.)

Wilson's love of nature comes through clearly in his autobiography and in the more engagingly written sections of *Sociobiology* and many of his later books.[34] It is a passion that cannot be overemphasized in explaining Wilson's personality and scholarship. Consider the opening paragraph of *Consilence*. Writing about his college days as a "beginning biologist" at the University of Alabama, Wilson notes fondly, "I had schooled myself in natural history with field guides carried in a satchel during solitary excursions into the woodlands and along the freshwater streams of my native state [Alabama]. I saw science, by which I meant (and in my heart I still mean) the study of ants, frogs, and snakes, as a wonderful way to stay outdoors."[35]

Is Wilson then just a "country boy" at heart? That is precisely the description of him offered by Salvador Luria, a molecular biologist and Nobel laureate at MIT.[36] Luria, another Marxist, was, like Chomsky, also a conspicuous nonmember of the Sociobiology Study Group. Luria surely could not have agreed with Wilson's politics, however it might ultimately be interpreted. But Luria apparently felt that the time and energy it would take to protest against *Sociobiology* could be better spent attacking really dangerous ideas, like those propounded by Jensen and Herrnstein. Luria also seems to have believed that Wilson's motives in writing *Sociobiology* were genuinely nonpolitical, and further, that Wilson was just naïve in not seeing that *Sociobiology* could be interpreted as supporting a conservative political agenda.[37] About his own political innocence in the 1970s Wilson has this to say in his autobiography: "In 1975 I was a political naïf: I knew almost nothing about Marxism as either a political belief or a mode of analysis, I had paid little attention to the dynamism of the activist left, and I had never heard of Science for the People. I was not even an intellectual in the European or New York–Cambridge sense."[38]

Wilson's political and intellectual "innocence" surely owed a good deal to his Southern Baptist upbringing. As a young child and an adolescent, Wilson took his religion very seriously, being "born again" at the age of fifteen, and otherwise taking part in the traditions of his faith. But by the time he had gone off to college, he had been seduced by nature and by the idea that the only way to understand its wonders was through the lens of evolutionary theory. Thus at an intellectual level Wilson's break with revealed religion was probably complete by the time he was in his late teens.[39] But Wilson, the new young naturalist, may not have been able to (or even have wanted to) shed completely the influences of his religious upbringing. A number of commentators have noted, for example, that Wilson's felicity with language—he does, after all, have two Pulitzer prizes—and in particular his sensitivity to the rhetorical dimensions of his writing may owe quite a bit to an evangelical tradition that emphasizes the need to touch both the minds *and* the hearts of individuals when spreading the Word.

But whatever the source of Wilson's rhetorical gifts, his religious upbringing seems also to have influenced his scientific worldview in at least one interesting way—though perhaps *influenced* is not quite the right word. It may simply be that there is an often unnoticed congeniality between the naturalist "spirit" that took hold of Wilson in his late teens and the natural theology teachings that are important to the worldviews of the dominant Western, and particularly Christian, religious traditions. Ullica Segerstråle does a good job of explaining this when she writes, "Traditionally, the naturalist spirit is connected to the idea of finding revelations of God's design in nature. For earlier Christian naturalists, God's law was not only to be found in the Bible but also in the Book of Nature. And elements of this lingered on in later generations; for instance, for Wilson's scientific 'grandfather,' [the eminent Harvard entomologist W. H.] Wheeler, the social insects still epitomized the moral virtue of hard work and co-operation and were upheld as a model also for human society."[40]

Wilson's "struggle" with religion may have been all too common, particularly for bright young adults who may have gotten too much religion too early. Wilson's problem was clearly with the authority of the priests, who claimed some special knowledge of God's plan and who insisted that this knowledge gave them alone the ability to say what the law was in the realm of ethics and morality. But Wilson's problem was not with those laws themselves, or at least not with the essence of those (Christian) laws. Anyone who spent any time in a Southern Baptist church would have been intimately familiar with this line from the New Testament: "Greater love hath no man than this, that a man lay down his life for his friends" (*John* 15:13). This is perhaps the animating principle of Christianity. There is no indication that Wilson ever rejected this idea. Quite the opposite. Defending this principle may actually be central to all his scientific endeavors. Thus it is absolutely striking that on the very first page of *Sociobiology*—a work that is said to represent the apotheosis of reductionism and scientific materialism—Wilson frames the "central theoretical problem of sociobiology" around the question "How can altruism, which by definition reduces personal fitness, possibly evolve by natural selection?"[41] Wilson poses this question even before he defines what he means by sociobiology.

But what is the love of which St. John was writing if not an extreme form of altruism? Wilson saw at a very early age the danger that science (and evolutionary theory in particular) posed to the principal ethical precept he had been raised to embrace. He seems to have spent a good part of his scientific life attempting to eliminate this danger by squaring scientific materialism (and, again, evolutionary theory) with the Christian concept of *agapē*. In one sense, the question of whether he succeeded at this is really beside the point. The more interesting question would seem to be, Why was he moved to make such an attempt in the first place? The obvious answer is that Wilson felt the power of science so strongly that he was unable to compartmentalize it and thus keep it away from the realm of morality or ethics. The point is not that Wilson felt a misplaced Oedipal desire to prove the priests were wrong.[42] The point is that Wilson felt an overwhelming desire to prove that the central ethical precepts in which he believed were based on something that was (to him) stronger than merely the authority of the priests.

There is, finally, one last aspect of Wilson's religious upbringing that I want to examine because of its possible influence on his science. It is an aspect that has not been very much discussed, as far as I can tell. It concerns the concept of original sin. If one believes, as any good Southern Baptist would, that humans are fallen beings, imperfectable in this (natural) world, one will probably tend to be unimpressed with utopian social schemes, like those peddled by Marxists. Such schemes have all been tried in the past, and they have failed. To be sure, there is something quite conservative about the embrace of original sin—even in its secular, philosophical sense as simply the imperfectability of human beings. What tempts me to think that Wilson may carry the baggage of original sin (or human imperfectability) is the extent to which this concept gets extended by Wilson even to science itself—even now, as we seem to stand on the threshold of the perfectibility of humankind through science. Thus in a striking passage at the conclusion of *Consilience* Wilson writes, "I predict that future generations will be genetically conservative. Other than the repair of disabling defects, they will resist hereditary change. They will do so in order to save the emotions and epigenetic rules of mental development, because these elements compose the physical soul of the species. . . . Why should a species give up the defining core of its existence, built by millions of years of biological trial and error?"[43] The language of this passage is scientific, but its tone is religious. We may have been given the wisdom to understand nature through science. But—Wilson seems to be saying—we were not given the wisdom, through science or through any other means, to perfect nature. Nor should we try, for such attempts would be like playing God.

On that note, I will conclude this examination of Wilson's politics and personality, and turn to a final question: Can sociobiology explain Edward O. Wilson? I raise this type of question whenever I examine the consistency of any grand theoretical system. Sociobiology is such a system and so is Marxism. As any number of critics have pointed out, one conspicuous problem with Marxism is that it has trouble explaining Karl Marx. Consider these famous first lines from Marx's "The German Ideology": "Hitherto men have constantly made up for themselves false conceptions about themselves, about what they are and what they ought to be. They have arranged their relationships according to their ideas of God, or normal man, etc. The phantoms of their brains have got out of their hands. They, the creators, have bowed down before their creations."[44] A more important *etcetera* than the one we find in this quotation has never been penned. In that etcetera lies Marx's first *rhetorical problem*.[45] Somehow Marx must convince his readers that, unlike all of the other systems men have ever dreamt of, the system *he* presents is not merely a phantom of his own brain. Somehow Marx must find a convincing way of exempting himself from his own critique. It is not clear that he succeeds at this, especially since the system he constructs looks strikingly like one that we would expect from a frustrated bourgeois intellectual.

Marx attempts to get around this problem by turning to the *material* basis of human society. He argues that by studying the material conditions of man's existence one can discover the historical laws that direct the progress of human history. As he famously explains, these laws prove that history ends in the communist utopia, a world in which "it is possible for me to do one thing today and another tomorrow, to hunt in the morning, fish in the afternoon, rear cattle in the evening, criticize after dinner, just as I have a mind, without ever becoming hunter, fisherman, cowherd or critic."[46] The key point here is that these historical laws are not mere phantoms of Marx's brain; they are real, scientific laws. They explain the development of history just as Darwin's theory of evolution explains the development of living beings.

Thus Marx solves his first rhetorical problem by grounding his theory in scientific inevitability. He is no myth-maker. He is simply demonstrating the way history must unfold. Unfortunately, this solution creates what might be called Marx's second rhetorical problem. If the communist world is inevitable, if it can no more be influenced through political action than can the laws of physics or the laws describing biological evolution, then it is not clear why anyone needs to struggle to bring it into existence. It is not clear why we need revolutionaries. To admit that Marxism needs revolutionaries, or even to admit that it must rely to some extent on politics, would seem to take Marxism out of the realm of science. But doing so would destroy the appeal of Marxism, for then it would become just another ideology. It would seem that both of Marx's rhetorical problems cannot be solved simultaneously. Either Marxism is a political project requiring the sustained commitment of individuals who freely *choose* to work toward revolutionary ends or it is a science that can guarantee those revolutionary ends as a matter of historical fact, thus eliminating the need for anyone to choose anything.

Scientists like Stephen Jay Gould face the first of Marx's rhetorical problems but not the second. As I have already noted, Gould's attempts to historicize science—his attempts to demonstrate how all past scientists (or at least those he does not agree with) were captives of their particular cultural or class prejudices—run the risk of faltering because of the self-referential paradox. Somehow Gould must convince us that *his* pronouncements on intelligence represent real science and not merely the ideological prejudices that he finds rampant in his predecessors' work.

On the other hand, scientists like Edward O. Wilson face the second of Marx's rhetorical problems, but not the first. Wilson is perfectly willing to concede that he is in no way exempt from his own pronouncements about the fundamentally empirical, and hence nontranscendental, quality of all aspects of all living things. He is perfectly willing to concede that, like everyone else, and indeed like every other living organism, he is simply a random (albeit unique) collection of amino acids and the proteins they code for. He is also more than willing to concede that *nothing* is inexplicable on materialist grounds. Thus he writes, "There is no apparent reason why [the brain sciences] cannot in time provide a material account of the emotions and ratiocination that compose spiritual thought."[47] God is dead, and it is science that killed him.

Wilson's problem is that all of his various concessions to the doctrine of scientific materialism undercut the persuasiveness of his own pleas, exhortations, and sermons that we try harder to save the planet. For example, in the closing pages of *Consilence* and throughout his 2002 book *The Future of Life*, Wilson offers an ardent appeal to the ethic of conservation and, in particular, to the idea that we do all we can to preserve as much of the biological diversity on Earth as possible even if that means we must somehow limit the number of humans on the planet. Wilson is clearly passionate about *biodiversity*. Doubtless his passion can be fully explained by a scientific analysis of the various chemicals in his brain and the way they interact one with another. Scientific materialism can readily provide such explanations, just as it can provide explanations for the passionate spirituality of other individuals.

But if science can do this it is hard to imagine toward whom, or rather toward what, Wilson is directing his passionate appeals for biodiversity. Surely he cannot be directing these appeals toward free beings capable of moral or ethical choice. Science has proven that such beings cannot exist. It has reduced what we may once have thought of as such beings to explainable biochemical units. And sociobiology has been instrumental in furthering this reductionistic approach—at least according to its critics.

This puts Wilson in a very difficult position. As the father of sociobiology he cannot abandon its claim to being a science, for then it would be nothing in his eyes. But Wilson is honest enough to admit that science is about reduction or it is just not science. He is also honest enough to admit that the more reductionistic science becomes the less room in our vocabulary there seems to be for concepts like morality and ethics. Wilson recognizes that this poses a problem for humans. He does not seek to evade this problem. Nor does he seek to provide a definitive solution to it. But in *Consilence* and elsewhere he does at least provide an important discussion of the complexity of the problem.

Other scientists, however, do seek to reconcile science with the belief that humans are capable of moral and ethical choice. Such scientists are critical of the work of sociobiologists precisely on the grounds that such work represents an approach to science that is too reductionistic. But is it possible to be critical of sociobiology on these grounds and still be a scientist? That is the question to which I now turn.

CHAPTER FOUR

Richard Lewontin
and His Colleagues Demur

A short nine years after E. O. Wilson published *Sociobiology: The New Synthesis*, Ronald Reagan won reelection to the presidency by the second largest electoral margin in the history of twentieth-century American presidential politics.[1] The year was 1984 and even a casual observer of the American scene at the time may well have concluded that it was not George Orwell's vision that had come to fruition in that fateful year but rather the vision of the nineteenth-century social philosopher Herbert Spencer. Indeed, Mario Cuomo, then a little-known governor from the state of New York, catapulted into the national spotlight that year by asserting, in his keynote speech at the Democratic National Convention, that the first four years of the Reagan presidency marked the sad triumph of "a kind of social Darwinism. Survival of the fittest."[2]

To their critics, both Social Darwinism and sociobiology were united in an unbending allegiance to the core philosophy of biological determinism. Many of these same critics were also quick to note that the success of Reagan in promoting his so-called Social Darwinist agenda and the success of Wilson in promoting sociobiology owed a great deal to the rhetorical gifts of each promoter. Thus by the early 1980s, many on the cultural and political left were calling for sustained efforts to develop a comprehensive and compelling response to the reigning ideology of the day—a response that would appeal not only to scientists but to the broader public as well.

One especially important result of these efforts was the publication, in 1984, of a book entitled *Not in Our Genes: Biology, Ideology, and Human Nature*.[3] The book was coauthored by Richard Lewontin, Steven Rose, and Leon Kamin. Lewontin needs no introduction at this point. Steven Rose is a neurobiologist and director of the Brain and Behaviour Research Group at Britain's Open University. He has sometimes been referred to as the "British Lewontin." In 2000 he coedited an important volume of scientific articles entitled *Alas, Poor Darwin: Arguments against Evolutionary Psychology*.[4] Leon Kamin is an honorary professor in the Department of Psychology at the University of Cape Town, South Africa. He came to national prominence in 1974 when, as a professor of psychology at Princeton, he published *The Science and Politics of IQ*, a book that forcefully critiques the work of psychometricians like Arthur Jensen and defends the position that there is no scientific evidence to show that intelligence is *to any degree* heritable.[5]

Not in Our Genes is an important book, written by three respected scientists, and directed primarily toward an educated audience of nonscientists. It generalizes and carries forth many of the arguments that the Sociobiology Study Group was making a decade earlier. As such, it constitutes nothing less than a frontal assault on the ideology of biological determinism. It also develops a sustained attack on the political agendas of Ronald Reagan, the New Right, and the capitalist-patriarchal-militaristic power structure in general. For these reasons at least it deserves our attention.

But *Not in Our Genes* is also a very curious book, in part because its authors are simultaneously making and confusing two related but distinct arguments. On one level, Lewontin, Rose, and Kamin are arguing that the various "theories" that the New Right and other conservatives allegedly use to buttress their particular ideological agendas—theories having to do with the heritability of intelligence or the biological differences between the sexes in areas like aggressiveness—are demonstrably wrong. Not only that, they assert that the wrongness of these theories can be demonstrated by science, and they set out to prove this, staying well within the established conventions of the scientific method. On the whole, this part of *Not in Our Genes* is at times very persuasive. Indeed, had Lewontin, Rose, and Kamin made only this type of argument they would have written a smaller, but ultimately more compelling and useful book than the one that actually got published.

But in addition to this argument, *Not in Our Genes* also advances an argument that takes place on a deeper, perhaps more philosophical level. This second broad argument is concerned not with the correctness of any specific scientific theory but with the overall wisdom of applying conventional scientific methods—particularly methods that seem to rely on the practice of reductionism—to the study of human societies or cultures. Throughout their book, Lewontin, Rose, and Kamin insist that the "reality" of the situation with respect to the connection between human nature and human society is always more complex than the story told by alleged reductionists and determinists like Wilson. At its core, this is an epistemological argument. The authors of *Not in Our Genes* are, at the very least, arguing for a vastly different understanding of science from the one that emerged, say, in the seventeenth century in Europe. They are not content to attack Reaganism and the New Right on a theory-by-theory, case-by-case basis using the conventions of science as it presently exists. They want to abolish the whole scientific and cultural apparatus that produces what they see as reductionistic and deterministic explanations for human societies and replace it with an apparatus or approach that produces what they call "dialectical" explanations for human society. In a moment, we shall come to a full discussion of what Lewontin, Rose, and Kamin mean by dialectical explanations. For now it is sufficient to point out that such explanations are, to put it charitably, a bit fuzzy. They tend to remind one of Heraclitus's statement that you cannot step into the same river twice. Thus Lewontin, Rose, and Kamin assert that "for dialectics the universe is unitary but always in change[;] the phenomena we can see at any instant are part of processes, processes with histories and futures whose paths are not uniquely determined by their constituent units." Hence, "according to the dialectic view, the properties of parts and wholes codetermine each other." From this it somehow follows that a scientific worldview based on dialectical explanations "abolishes the antitheses of reductionism and dualism; of nature/nurture or of heredity/environment; of a world in stasis whose components interact in fixed and limited ways, indeed in which change is possible only along fixed and previously definable pathways."[6] While it may not be immediately obvious exactly what the authors of *Not in Our Genes* are trying to get at here, it does seem that their scientific worldview is fundamentally different from and, in their eyes at least, superior to any explanation that can be provided by what we now think of as science.

But therein lies the problem. By making this claim, Lewontin, Rose, and Kamin have painted themselves into a corner. Simply put, the two levels of argument that run throughout *Not in Our Genes* are in fairly strong tension with one another. By seeming to attack the scientific method as such, and by seeking to replace it with a method that produces dialectical explanations for phenomena, Lewontin, Rose, and Kamin may be undercutting the persuasiveness of the scientific arguments that they make against the very theories that they do not like. In the end, the reader is left to wonder just what explanatory paradigm the authors are offering, and whether it does in fact offer us more promise than conventional scientific methods in our quest to defeat the forces of economic and social injustice.

To get a better handle on the precise nature of the tension that I have said runs throughout the pages of *Not in Our Genes*, and to see better how this tension works to undercut the book's overall persuasiveness, I want to focus briefly on an important issue that Lewontin, Rose, and Kamin discuss at some length. The issue is the heritability of intelligence. As we have seen in chapters one and two, intelligence has been and continues to be a hot-button issue in any society, but particularly in advanced industrial societies in which we find a fairly high degree of economic stratification. To appreciate the significance of the issue involving the heritability of intelligence, one needs to understand the precise meaning of the term *heritability* as it is used in the biological sciences.

Heritability tells us something about individual physical characteristics or traits (any characteristic or trait that can be measured, in fact) *within a sufficiently large population of individuals*. Consider the population of American adult males. Within this large population one finds individuals of various weights. Some may weigh as little as one hundred pounds while others may weigh as much as three or four hundred pounds, with the average somewhere between these extremes. Weight is obviously a physical characteristic that can be easily measured. It is also a characteristic that is influenced both by an individual's genes and by his environment.

If you are curious about what percentage of the total variance of a given trait in a given population is attributable to genetic factors, you are curious about *heritability*. Remember, heritability is a concept that applies *only* to populations of individuals, not to individuals themselves. By definition, the heritability of a given trait in a given population is the *ratio* of the genetic variation of the trait in the given population to the overall variation of the trait in the population, expressed as a percentage. The overall variation of a given trait in a given population is, in turn, a combination of genetic and environmental variation.

If *all* of the variance of a given trait in a given population is attributable to genetic factors, the trait's heritability (*in that population*) is said to be 100 percent. In contrast, if *none* of the variance of a given trait in a given population is attributable to genetic factors, the trait's heritability is said to be zero. There are some fairly obvious and noncontroversial examples of traits with zero heritability. Consider the variation among native languages of individuals in America. As we will see in chapter eight, all human languages are produced using the same basic genetic equipment. Although there may be more than one mental module for language acquisition *in general*, there is *not* a mental module for English language acquisition and a separate mental module for Japanese language acquisition. In fact, every human infant is born with the ability to learn any language as well *and as easily* as any other. Hence, the genetic ability of any given American infant to learn English does not vary from his or her genetic ability to learn Spanish. Thus the fact that any given American's *native* language is Spanish as opposed to English is due entirely to that individual's environment—generally whether he or she grew up in a home in which Spanish as opposed to English was the family language. The same could be said for any given American's religious affiliation. The fact that any given American is a practicing Jew as opposed to a practicing Christian is due entirely to the individual's environment. But having said that, I cannot resist pointing out that, although an individual American's religious affiliation will display a heritability of zero, there is some tantalizing evidence that an individual American's *overall level of religiosity* (whether he or she is a passionate member of any faith or merely a weekend visitor to a church or synagogue) may be influenced by genetic factors and thus may be *to some degree* heritable. This might be at least partially explainable by the conjecture that one's level of religiosity is tied to one's overall temperament, and that one's temperament is affected by genetic factors, thus making it to some degree heritable.[7]

But *is* temperament heritable *to any degree*? This is not an uncontroversial question. Indeed, it turns out that while there may be many noncontroversial examples of traits within a given population that display a heritability of zero, it is considerably more difficult to find noncontroversial examples of traits within a given population that display a heritability approaching 100 percent. Even eye color is not 100 percent heritable, since it can be affected by environmental factors. Traits like height and weight, not to mention intelligence and temperament, are obviously not 100 percent heritable either.

At this point in our discussion, I imagine that were he still alive Stephen Jay Gould would want to emphasize the essential "unreality" of heritability. Gould might emphasize this by way of the following warning: *Remember, heritability is simply a statistical construct. It is not real in the sense that a gene is real. It can be altered by political arrangements or statistical manipulation.* This is an important point and one that cannot be stressed too often. A critical consequence of this point is that you can change the heritability of a given trait simply by changing the population under examination.

To better understand this, consider two different populations: one composed of one hundred males selected randomly from the senior class at the Virginia Military Institute, and the other composed of one hundred males selected randomly from the senior class at the Virginia Polytechnic Institute. For those unfamiliar with higher education in the Commonwealth of Virginia, the Virginia Military Institute is a state-funded military college located in the town of Lexington. Under court order it started admitting women in 1997 and has only a small percentage of women in its student body. The Virginia Polytechnic Institute is a larger, but also state-funded, university located about eighty miles to the south of VMI in the town of Blacksburg. It is what most people would consider a "regular" college, with an emphasis on science and engineering.

With respect to these two sample populations it is almost certainly the case that the physical trait of body weight is more highly heritable within the VMI sample than within the VPI sample, even though both institutions probably draw students from roughly the same large gene pool. Do you object to that last assumption about the similarity of the gene pool of students at both institutions? Fine. Assume that starting next year admission to both VMI and VPI were forced and strictly random. This could be done by having the names of all eighteen-year-olds in America thrown into a big pool from which the entering classes at both institutions were drawn, with attendance being compulsory if one's name were selected for one or the other institution. Under these admittedly bizarre circumstances, in four years we could be even *more* certain than we are now that within random samples of one hundred male seniors drawn from each institution body weight would be more highly heritable in the VMI sample than in the VPI sample. The reason is that the environment at the Virginia Military Institute is more nearly equal for each student than the environment at the Virginia Polytechnic Institute. Every student at VMI eats roughly the same food, is made to exercise roughly the same amount of time per day and in roughly the same way, gets roughly the same amount of sleep, has access to roughly the same level of medical care, and so forth. On the other hand, all of these environmental factors vary considerably among students at VPI.[8] Because whatever variation we observe among seniors at each institution must (obviously) be the result of some factor that can vary, and because we have assumed that the environment varies less at VMI than VPI, the variance among the body weight of seniors at VMI must be due more to genetic factors than the variance among the body weight of seniors at VPI is due to genetic factors. Hence, among seniors at VMI body weight will be more highly heritable than it will be among seniors at VPI.

Another way of demonstrating the same point is to suppose that you took both of the above one hundred student samples and plotted the body weight of each sample on a separate sheet of paper. The easiest way to do this would be to take a sufficiently wide sheet of paper and draw a line at the bottom that was, say, forty centimeters (roughly sixteen inches) long. Put a small vertical dash at the extreme left edge of the line, and under the dash write 100—for those few, if any, students who weigh only one hundred pounds. Now proceed one centimeter to the right and mark off another dash. Under this dash write 105. Continue in this manner across the entire line and you will have forty-one dashes that delineate weights from 100 to 300 pounds, in five-pound intervals. Now take either of your two one-hundred-student samples. For each student in the sample place a dot on your paper above the dash that most closely corresponds to that student's weight. If two or more students have the same weight, place their dots, one above the other, in a vertical line above the corresponding dash. Finally, step back from your paper. The shape that you would see should approximate that of a bell. If you connect the dots that form the outline of the bell, you will have a bell-shaped curve.

It is almost certain that the curve corresponding to the VMI students would be more pointed and narrowly grouped around its center than the curve corresponding to VPI students. Now suppose that you select one student from the same tail end of each curve—say, the "heavy" end of the curve. Because body weight is more highly heritable at VMI than VPI, you could be more confident that one or both of the biological parents of the VMI student were "heavy" than you could be that one or both of the parents of the VPI student were "heavy."

This purely statistical point about heritability can be connected to intelligence in an obvious way. Recall the claim made in chapter two by Richard Herrnstein that, as the environment in America becomes more equal for everyone, the heritability of intelligence will necessarily increase. This assumes, of course, that the genetic factors that go into making up a person's intelligence—factors that everyone agrees are real in a physical sense—can vary at all from individual to individual. This assumes, in other words, that the heritability of intelligence is not zero.

So what exactly is the magic number for the heritability of intelligence in America? This is a question over which there is much heated debate. One point that is not open to debate, however, is that in America intelligence—at least as best we can measure it—does seem to vary considerably from individual to individual. How can we explain this variance?

As Lewontin, Rose, and Kamin point out, an explanation that establishes that intelligence is to a high degree heritable would seem to suit "Reagan conservatives" quite well, especially if one believed (and this does not seem like much of a stretch) that highly intelligent people are more likely than less intelligent people to succeed in society by, for example, doing well in school, or getting a high-paying job. In fact, an explanation that "proved" intelligence to be highly heritable might allow (or even compel) one to believe that the children of rich individuals do better in society not because their parents pass on to them hefty trust funds, but rather, because their parents pass on to them genes for high intelligence.

But is it *scientifically* true that intelligence is highly heritable? Much is riding on the answer to this question. The authors of *Not in Our Genes* want to demonstrate that this is not true, and they set out to do so by analyzing and critiquing the scientific evidence that tends to support this claim, and then by presenting scientific evidence to the contrary. At this, they are most persuasive. To see why, we need to examine their examination of the scientific evidence on both sides of this claim.

What scientific evidence tends to support the claim that intelligence is highly heritable? It is generally acknowledged, by all sides in the debate, that in "normal" households a child's intelligence, as measured by standard IQ tests, tends to correlate strongly with his or her parent's intelligence, again as measured by standard IQ tests. Recall from our discussion in chapter two that a correlation is the degree of similarity between two variables. In this case the two variables are the child's intelligence and either the mother or the father's intelligence. Obviously the intelligence of the mother and father, as measured by standard IQ tests, will rarely be exactly the same. But, as it turns out, there is also a very high correlation between the intelligence of the parents of any given child, perhaps because people of similar intelligence tend to marry—a trend that may be increasing.[9]

But this high correlation between the levels of intelligence of parents and their biological children tells us nothing about the heritability of intelligence as such. It may be the case that this high correlation is entirely the result of the similar household environments that parents share with their children. Or it may be the case that it is entirely the result of the similar upbringings that biologically related parents and children share from generation to generation. Or finally it may be the case that this high correlation is *entirely* the result of genetic factors that parents share with their biological children. How are we to decide among these various explanations?

As it turns out, early in the twentieth century, two researchers, B. S. Burks and A. M. Leahy, set out independently to answer the question I have just posed. Both researchers designed essentially the same experiment. Their "classical" design is described by Lewontin, Rose, and Kamin.[10] First, the researchers set out to find a sufficiently large number of adopted children who were being raised by adoptive parents. Next, the researchers set out to "match" the households in which these children were raised with an equal number of identical households in which children were being raised by their biological parents. Finally, the researchers determined the correlation between the IQs of the children and their parents in each separate group. What they found was that the IQs of children and parents in the first group (the group composed of adopted children) showed very little positive correlation. The figure was around 0.15. On the other hand, the IQs of children and parents in the second group (the group composed of children and their biological parents) showed a significantly higher positive correlation. The figure for this group was around 0.48. The researchers argued, not unpersuasively, that whatever correlation was found in the second group must necessarily be the result of some combination of environmental and genetic factors. Because the correlation in the second group was so much higher than that in the first group, and because it was assumed that the two groups were identically matched household for household, the researchers concluded that although changeable environmental factors may play a small part in determining a person's IQ, genetic factors play a much greater part. Hence, intelligence is highly heritable.

Lewontin, Rose, and Kamin need at least to find fault with this "classical" study design and its results in order to advance the claim that intelligence is not significantly heritable. They begin by probing the weakest part of the study design: the assumption that the two groups were matched household for household. They make much of what they consider the "impossible requirement of matching adoptive and biological families."[11] They point out, for example, that although Burks and Leahy attempted to match households on the basis of standard demographic characteristics like the occupations and educational levels of the parents and the quality of the neighborhoods in which these parents lived, the researchers did not (or could not) match households on the basis of some other fairly obvious characteristics like the age of the parents and even household income. Beyond that, Lewontin, Rose, and Kamin point out that Burks and Leahy failed to match households on the basis of perhaps the most significant environmental factor in any child's life: whether the child is wanted or not. The researchers could not do this because, obviously, no adoptive parents did not want to have children.[12]

But if the environment of all adopted children is the same (or nearly the same), one cannot draw any meaningful correlation between that environment (or any aspect of that environment, including the adoptive parents' intelligence) and any variable, like the intelligence of adopted children. Remember, *a correlation can exist only between variables*. Lewontin, Rose, and Kamin are arguing, based on some fairly convincing evidence, that adoptive households show little variability, certainly much less variability than households in which parents raise only their biological children. In fact, when compared to nonadoptive parents, adoptive parents tend to be *uniformly* more intelligent, wealthier, older, and more experienced at raising children, since they tend to have biological children before they adopt children. This being the case, we would expect the correlation between the intelligence of adoptive parents and the intelligence of their adopted children to be relatively low. Indeed, this expectation would be even greater if adoption agencies tended randomly to place children in adoptive families and/or if adoptive families (as a group) tended to adopt children from a wide range of backgrounds, including children of various races, ethnicities, and nationalities. Again, if this were the case, then the correlation between the intelligence of parents and their adopted children would be low because only one variable (the intelligence of the adopted children) was varying. To take an analogy, recall our earlier case of the obsessive roulette player who always plays only one number, even though the size of his wager varies from spin to spin. We noted that there would be a zero correlation between the size of this individual's wager and the chance that his number would come up, precisely because the chance that his number would come up does not vary. Similarly, if the intelligence of adoptive parents does not tend to vary that much, but the intelligence of the children they adopt does, we would notice a low correlation between the intelligence of the adoptive parents and the intelligence of their children. Once again, according to the authors of *Not in Our Genes*, the *basic* problem here lies in the impossible requirement of matching adoptive and biological families.

But Lewontin, Rose, and Kamin do quickly point out that there is a fairly obvious and elegant way around this problem. The solution is in fact contained in the above paragraph. Suppose we examine only the households of those adoptive parents who also have biological children of their own. For any given such household, adoptive and biological families would (obviously) be identical. We could then correlate the intelligence of one of the adoptive parents (say, the mother) in any given such household *twice*: first we could correlate the intelligence of the mother with the intelligence of her biological child and, second, we could correlate the intelligence of *the same woman* with the intelligence of her adopted child.

In fact, as the authors of *Not in Our Genes* note, at least two studies have been done using this methodology. Both studies were conducted in the late 1970s, one in Minnesota and the other in Texas. As Lewontin, Rose, and Kamin also note, "The investigators in each case were behavioral geneticists who clearly expected to discover evidence supporting a high heritability of IQ."[13] As the authors of *Not in Our Genes* clearly imply, we should have all the more confidence in these studies if the results were *not* what the investigators expected. Indeed, the results seem to show that IQ is *not* heritable. This is how Lewontin, Rose, and Kamin report and analyze the findings:

> The results for mother-child pairings in both studies are as follows: The same mother's IQ, remember, has been correlated with the IQ of her adopted and of her biological child. There is no significant difference between the two correlations. In Texas the mother was a trifle more highly correlated with her adopted child, and in Minnesota with her biological child. The Minnesota study, it might be noted, was based upon transracial adoptions. That is, in almost all cases the mother and her biological child were both white, while her adopted child was black. The child's race, like its adoptive status, had no effect on the degree of parent-child resemblance in IQ. These results appear to inflict fatal damage to the notion that IQ is highly heritable. Children reared by the same mother resemble her in IQ to the same degree, whether or not they share her genes.[14]

Whether the damage that these results inflict on the notion that IQ is highly heritable is in fact *fatal* is perhaps questionable. But the damage is surely considerable. On the other hand, there is more than one way to skin the IQ-heritability cat. The two studies mentioned above measured the heritability of

IQ by taking two *genetically different* individuals (two children who were not genetically related) and placing them in an *identical environment* (the household of the biological mother of one of the two children). It was then found that the genetic difference between these two individuals made no measurable difference in the correlation of these individuals' intelligence with the intelligence of their mother. Thus it was concluded that genetic variation accounts for a relatively small amount of the variation one finds in intelligence.

But suppose one could take two *genetically identical* individuals and place each in a *different environment*. One could then test to see how traits like intelligence are affected when the environment is varied but genes are held constant. Exactly such studies have been done on monozygotic twins who have been reared apart (MZAs, as they are sometimes called in the literature). Monozygotic twins are the product of the same fertilized egg (zygote), which splits in two, ultimately forming two separate, but genetically identical, individuals. Obviously, most monozygotic twins are reared in the same family. But some are, for whatever reasons, reared apart in different environments. A study published in the 12 October 1990 issue of *Science* examined fifty-six pairs of monozygotic twins reared apart. These pairs had been observed since 1979. The study, conducted by Thomas Bouchard and others, looked at the similarity of these pairs with respect to several psychological traits. The results concerning IQ were striking. As the authors report, "The IQs of the adult MZA twins assessed with various instruments in four independent studies correlate about 0.70, indicating that about 70% of the observed variation in IQ in this population can be attributed to genetic variation."[15] This is an enormously significant correlation, well above the correlation of between 0.40 and 0.50 for siblings who are not monozygotic twins but who are reared *together*.

The obvious question raised by this exceedingly high correlation concerns the issue of *correlated placement*: "Were the twins' adoptive homes selected to be similar in trait-relevant features which, in turn, induced psychological similarity?"[16] Bouchard and his coauthors examine this question at great length. They note that, for a variety of factors (including the educational levels of the adoptive parents, the socioeconomic status of the adoptive families, the access of the adopted children to scientific and technical literature within the home, and so forth) the amount of the overall correlation (of 0.70) that can be explained by even the most significant of these environmental factors is vanishing small, at around 0.03.[17] The claim that the high correlation found in this study is largely the result of genetic, rather than environmental, factors receives support from another interesting finding. As Bouchard and his co-authors write, "adult MZ twins are about equally similar on most physiological and psychological traits [including intelligence], *regardless of rearing status.*"[18] Extrapolating from all of this and more, the authors conclude, "The present findings . . . indicate that, in the current environments of the broad middle-class, in industrial societies, two-thirds of the observed variance of IQ can be traced to genetic variation."[19] Thus, intelligence is highly heritable after all.

It should be noted that Lewontin, Rose, and Kamin obviously have nothing to say about the specific monozygotic twin study I have just examined, since it was published after their book was written. Nor, really, do they have much to say about twin studies in general. They do concede that most studies of monozygotic twins reared *together* show a remarkably high correlation in intelligence of between 0.70 and 0.90 for twin pairs. And they also concede that this correlation is higher than IQ correlations for pairs of nonmonozygotic twins reared together or than correlations for other pairs of siblings reared together. They put both of these latter two correlations in the relatively high 0.50 to 0.70 range. But Lewontin, Rose, and Kamin attempt to explain the high IQ correlations for monozgyotic twins reared together by insisting that such twins are in fact treated nearly identically by parents, teachers, and other siblings. Even so, this explanation in no way addresses the study design of Bouchard and his coauthors—a study design that seems to defeat quite conclusively the objection that the high IQ correlations between monozygotic twin pairs is due to the fact that both twins are treated exactly the same by those with whom they interact. Thus it is not clear how much weight we should give to Lewontin, Rose, and Kamin's conclusion that "twin studies as a whole, then, cannot be taken as evidence for the heritability of IQ."[20]

It should also be noted that Herrnstein and Murray rely extensively in *The Bell Curve* on twin studies—and specifically on the study by Bouchard and his colleagues—to substantiate the claim that IQ is highly heritable. Based on their assessment of all the relevant studies, Herrnstein and Murray adopt what seems like a moderate sounding position when they write, "We are content . . . to say that the heritability of IQ falls somewhere within a broad range and that, for purposes of our discussion, a value of 0.6 ± 0.2 does no violence to any of the competent and responsible recent estimates. The range of 0.4 to 0.8 includes virtually all recent (since 1980) estimates—competent, responsible, or otherwise."[21] The low end of this range, 0.40, may not be as high as the 0.70 figure that Bouchard and his coauthors advance, but it is certainly not zero. Hence, if even this extremely conservative estimate is true, that fact would, one might say, appear to inflict fatal damage to the notion that IQ is *not* highly (or at least *significantly*) heritable.

Is IQ then significantly heritable or not? In answering this question I am tempted to respond with the four words that often conclude many a social science article: "more research is needed." But saying this would seem to suggest that, with more scientific research, there is *hope* that a satisfying answer to this question may indeed be found. It is precisely this hope that Messrs. Lewontin, Rose, and Kamin seek to dash in their book. Not only do they seek to dash this particular hope; they seek to dash any hope that science as it is presently constructed will be able to provide us answers to any questions regarding the relationship of biology to culture, or of nature to nurture. Presumably they seek to destroy these hopes because they believe that such hopes are of a piece with the empty promises offered by biological determinists and scientific reductionists. In short, Lewontin, Rose, and Kamin believe that false hopes are worse than no hope at all.

They also believe that by the mid-1980s those false hopes were peddled most beguilingly by sociobiologists. Thus in the penultimate chapter of their book they seek to unmask sociobiology's hidden ideological agenda. "Sociobiology," they write,

> is a reductionist, biological determinist explanation of human existence. Its adherents claim, first, that the details of present and past social arrangements are the inevitable manifestations of the specific action of genes. Second, they argue that the particular genes that lie at the basis of human society have been selected in evolution because the traits they determine result in higher reproductive fitness of the individuals that carry them. The academic and popular appeal of sociobiology flows directly from its simple reductionist program and its claim that human society as we know it is both inevitable and the result of an adaptive process.[22]

For Lewontin, Rose, and Kamin it seems that the *basic* problem with sociobiology—indeed the basic problem with science in general—is that it simply cannot make good on its grandiose promises to provide real knowledge of human societies or cultures. Indeed, the authors of *Not in Our Genes* never tire of pointing out what they assert to be the "evident failure of reductionism as a methodological program in the study of society."[23] But even assuming that this failure is as evident as they insist, one may still justifiably ask, What *alternative* to reductionism can these three scientists offer for the study of society? The authors of *Not in Our Genes* realize—just as did Stephen Jay Gould at the conclusion of *The Mismeasure of Man*—that unless they can provide an answer to this question they risk being dismissed by the general populace as what Spiro T. Agnew called "nattering nabobs of negativism." Agnew's reference was to the members of the "elite" media who could not seem to find anything nice to say about President Nixon. But Lewontin, Rose, and Kamin must have been worried that something like that description— albeit probably one less alliterative—could have been applied to them. Hence, in place of the biological determinism and scientific reductionism that they see as inherent in the work of Edward O. Wilson, Richard Dawkins, and other sociobiologists, the authors of *Not in Our Genes* advance an approach toward knowledge in general and biology and culture in particular that they call "dialectical."

At first glance, a dialectical approach to the study of anything may look superior to its (nondialectical?) alternative. But the more Lewontin, Rose, and Kamin attempt to describe their dialectical alternative the more it seems that what they are really after is not just an alternative approach to the doing of science, but, more generally, an alternative approach to the conceptualization of knowledge. This is a tall order for any book. It is not at all clear that they come close to realizing their goal. Indeed, about the only point that does become clear throughout *Not in Our Genes* is that whatever else the dialectical approach may be, it is not, under any circumstances whatsoever, reductionis-tic or deterministic. It is not *biologically reductionistic*, because it does not reduce human beings to mere collections of genes that then determine be-havior. Nor is it *culturally reductionistic*, because it does not posit that humans are blank slates on which society writes rules which determine behavior. Because they want to be especially clear on what their approach is not, Lewon-tin, Rose, and Kamin write, "The contrast between biological and cultural determinists is a manifestation of the nature-nurture controversy that has plagued biology, psychology, and sociology since the early part of the nine-teenth century. Either nature plays a determining role in producing the simi-larity and differences among human beings, or it does not, in which case, what is left but nurture? We reject this dichotomy."[24]

Very well. But again the question can justifiably be asked, What is the alternative to this dichotomy? Perhaps a suitable alternative can be found in an approach that might be called *interactionism*. According to Lewontin, Rose, and Kamin, interactionism holds that "it is neither the genes nor the environment that determines an organism but a unique interaction between them."[25] This does indeed sound promising. Lewontin, Rose, and Kamin even concede that "interactionism is the beginning of wisdom."[26] But then that old negativism creeps in again. Hence, the authors of *Not in Our Genes* conclude,

> Interactionism, while a step in the right direction, is flawed as a mode of explanation of human social life. It carries with it two basic assumptions that it has in common with more vulgar determinisms and that prevent its solving the problem of society. First, it supposes the alienation of organism and environment, drawing a clean line between them and supposing that environment makes organism, while forgetting that organism makes environment. Second, it accepts the ontological priority of the individual over the collectivity and therefore of the epistemological sufficiency of the explanation of individual development for the explanation of social organization. Interactionism implies that if only we could know the norms of reaction of all living human genotypes and the environments in which they find themselves we would understand society. But in fact we would not.[27]

Here we get a clue about the possible reason for Lewontin, Rose, and Kamin's negativism. Perhaps they reject so many scientific approaches because their view of these approaches is either too shallow, or too highly caricatured, for *anyone* to embrace them. Interactionism serves as an example of this problem. One does wonder what these authors mean when they assert that interactionism forgets that "organism makes environment." Far from forgetting this, interactionism is grounded upon this idea. It is precisely what is meant by *inter*actionism. One also wonders about the exact nature of the criticism implied by the charge that interactionism "accepts the ontological priority of the individual over the collectivity." Perhaps interactionism is not Marxist enough. Similarly, the charge that interactionism is flawed because it accepts the epistemological sufficiency of the explanation of individual development for the explanation of social organization once again misses the whole point of interactionism. (The point is, however, missed with the help of some philosophically sophisticated sounding language.) Writing an exasperated review of *Not in Our Genes*, Richard Dawkins, whom Lewontin, Rose, and Kamin describe as an archreductionist, points to passages like the one about interactionism quoted above and concludes: "This sort of writing appears to be intended to communicate nothing. Is it intended to impress, while putting down smoke to conceal the fact that nothing is being said?"[28] Dawkins's question is of course rhetorical.

Finally, after critiquing a number of alternative approaches to the study of society, Lewontin, Rose, and Kamin appear to offer what they seem to believe is a new approach to science. What they offer is instead an anticlimax to their book:

> We would insist on the unitary ontological nature of a material world. . . . The biological and the social are neither separable, nor antithetical, nor alternative, but complementary. All causes of the behavior of organisms, in the temporal sense to which we should restrict the term *cause*, are simultaneously both social and biological, as they are all amenable to analysis at many levels. All human phenomena are simultaneously social and biological, just as they are simultaneously chemical and physical. Holistic and reductionist accounts of phenomena are not "causes" of those phenomena but merely "descriptions" of them at particular levels, in particular scientific languages. The language to be used at any time is contingent on the purposes of the description.[29]

What is striking about this passage—to me at least—is how nonrevolutionary it sounds, and indeed how thoroughly pragmatic is its approach to science. (Marxists are not generally known for their pragmatism.) Lewontin, Rose, and Kamin seem to be saying that we ought not to view scientific descriptions of ourselves as privileged over any other descriptions we could create. Instead, we should view these descriptions simply as tools we can use when we want to predict and control the natural world and things in it— things like human beings. Sometimes these descriptions work well and are hence useful. But even when they are useful they are just that, and nothing more. Thus we should see science as one more tool we can use to help us cope with our world.

If this is what is involved in a "dialectical approach" to science, it is hard to see what all the fuss is about. To be sure, this is good counsel. But, as I have said, it is not revolutionary. Nor is it antithetical to what most scientists actually do when they engage in science. Indeed, Patrick Bateson, a professor of ethology at Cambridge University, a contributor to Rose's recent book *Alas, Poor Darwin*, and generally a sympathetic audience for those who would attack sociobiology, can manage at best only a lukewarm review of *Not in Our*

Genes. In characterizing its authors' dialectical approach to the study of society, Bateson writes,

> Their concluding discussion of the interplay between many different influences, reciprocity between the individual and the social and physical environment, and the emergence of properties in systems is well done. But it is not new and is not made any more so as a result of their slightly grumpy attempts to distance themselves from all others who espouse "interactionism." Possibly some real people hold the limited interactionist view as it is described in this book, but since they are not mentioned by name it is hard to be sure. As far as I am concerned, the position finally reached by Rose, Kamin and Lewontin is correct, shared by many others and, dare I say it, rather moderate.[30]

Moderate or not, the most interesting aspect of the dialectical approach that these authors champion is that it never really addresses the critical issue concerning scientific reductionism. Even if reductionistic accounts of human behavior are not "causes" of these behaviors but merely "descriptions" of them at particular levels, if these descriptions work very well at predicting human behavior one does begin to wonder how anything like a concept of free will can be sustained. *This* is the problem of reductionism. But Lewontin, Rose, and Kamin's discussion of the dialectical approach to the study of society, however philosophical it may sound, simply does not address this problem. Indeed, the best they can offer is this concluding paragraph of *Not in Our Genes*:

> What characterizes human development and actions is that they are the consequence of an immense array of interacting and intersecting causes. Our actions are not at random or independent with respect to the totality of those causes as an interacting system, for we are material beings in a causal world. But to the extent that they are free, our actions are independent of any one or even a small subset of those multiple paths of causation: that is the precise meaning of freedom in a causal world. When, on the contrary, our actions are predominantly constrained by a single cause, like the train on the track, the prisoner in his cell, the poor person in her poverty, we are no longer free. For biological determinists we are unfree because our lives are strongly constrained by a relatively small number of internal causes, the genes for specific behaviors or for predispositions to these behaviors. But this misses the essence of the difference between human biology and that of other organisms. Our brains, hands, and tongues have made us independent of many single major features of the external world. Our biology has made us into creatures

who are constantly re-creating our own psychic and material environments, and whose individual lives are the outcomes of an extraordinary multiplicity of intersecting causal pathways. Thus, it is our biology that makes us free.[31]

In other words, our *human* biological complexity gives rise to culture and in so doing defeats reductionism, thus guaranteeing our freedom. About the only thing one could say for sure of this position is that in advancing it Lewontin, Rose, and Kamin seem to have made their peace with natural science. But one fears that the authors of *Not in Our Genes* may simply be confusing freedom with ignorance. Even so, the value of their book is not to be found in the chapters that grapple with the deep philosophical questions of free will and determinism. Lewontin, Rose, and Kamin are no better (but also no worse) than any other intellectuals at shedding light on these mysteries. Rather, the value of *Not in Our Genes* can be found in the chapters—like the one on intelligence—that subject science to precise and demanding critique, using its own methods. In doing this, Lewontin, Rose, and Kamin are doing the necessary work of keeping scientists honest. This is all the more important in the twenty-first century as scientific theories seek to explain the human mind itself in greater and greater detail—and as they become more successful at this task. Evolutionary psychology is now the most prominent and perhaps the most promising of these scientific theories. The next part of this book is devoted to an explanation of its central principles.

PART TWO

The Blind Watchmaker Meets the Scatterbrained Computer Programmer

Evolutionary psychology brings together two scientific revolutions. One is the cognitive revolution of the 1950s and 1960s, which explains the mechanics of thought and emotion in terms of information and computation. The other is the revolution in evolutionary biology of the 1960s and 1970s, which explains the complex adaptive design of living things in terms of selection among replicators. The two ideas make a powerful combination. Cognitive science helps us to understand how a mind is possible and what kind of mind we have. Evolutionary biology helps us to understand *why* we have the kind of mind we have.

—Steven Pinker
How the Mind Works, 1997

CHAPTER FIVE

Nature's "Very Special Way"

The title of this part of the book layers multiple allusions. Let me begin by trying to unpack some of those allusions. The blind watchmaker appears as the title character of a 1986 book by Richard Dawkins.[1] Significantly, the subtitle of Dawkins's book is "Why the Evidence of Evolution Reveals a Universe without Design." Also, significantly, Dawkins's main title—*The Blind Watchmaker*—is itself an allusion to a famous argument *against* evolution advanced by William Paley, the great eighteenth-century theologian.

Paley's argument proceeds by way of an analogy. He begins by noting that if I am walking along a path and stub my toe on a stone, I have no great occasion to wonder how the stone came to be there. If someone presses me for an answer, I might say that for all I know or care the stone has been there for as long as the earth itself existed. But now suppose I am walking along the same path and look down to see a watch on the ground. I immediately ask myself how it came to be there and I find that the answer I gave for the stone simply will not suffice. There is no way I can believe that a watch, with its intricate design so obviously intended for some purpose, is itself the product of the same random *physical* forces that I can believe "constructed" the stone. Because the watch has a complex design and an evident purpose, there must be a watchmaker who designed it for that purpose. Paley then completes the analogy by insisting on the self-evident truth that *works of nature*—from plants and animals to humans—are more like watches than stones. This is how he puts it in his 1802 book *Natural Theology*: "Every indication of contrivance, every manifestation of design, which existed in the watch, exists in the works of nature; with the difference, on the side of nature, of being greater or more, and that in a degree which exceeds all computation."[2]

Paley's point was simply that where there is a design there must be a designer and where there is a very good design there must be a very good designer. But more than that, the designer must be *separate* from the thing designed. Even the most sophisticated watch certainly cannot design itself. Of course, today we do have computers that design other computers and programs that write other programs. But Paley would not be concerned. There still needs to be a first designer or a first programmer that is not itself a computer or a program.

Interestingly, in his *Critique of Judgment* the philosopher Immanuel Kant—another eighteenth-century figure of note—uses the same example of a watch to help make Paley's point. Kant notes, as I just did, that inanimate matter cannot design, much less animate, itself. "That is," Kant says, "the reason why the cause that produced the watch *and its form* does not lie in nature (the nature of this material), but lies outside nature and in a being who can act according to *the ideas of a whole* that he can produce through his causality."[3] Again, for Kant, as for Paley, design is the key. Watches may, in some sense, be produced from the metals of the earth, but the form of the watch—its design—is produced by a watchmaker. Hence, when we see design we know that there must be an intelligence at work. Indeed, we *must* assume this, but *only* when we see design. Thus Kant notes, quite correctly, that unlike the physical sciences, which can simply study the effects of various forces on various bodies without troubling themselves any further, the biological sciences must assume that the things they study are more than simply a random collection of particles. They must assume that these things have an *intrinsic* design and that each part of the thing is somehow related to this design. Additionally, they must assume that these things are all somehow part of some plan *extrinsic* to the thing itself, a plan that Kant will only go so far as to call an "analogue of life."[4] To drive home his argument Kant draws our attention to the "familiar fact that those who dissect plants and animals in order to investigate their structure" *must* follow "the maxim that nothing in such a creature is *gratuitous*. They appeal to it [this maxim] just as they appeal to the principle of universal natural science—viz., that *nothing* happens *by chance*. Indeed, they can no more give up that teleological principle than they can this universal physical principle."[5]

But if in order to understand works of nature we must assume that they have a designer, and if this designer cannot itself be a work of nature (for then we would have to explain who or what designed it), where should we look for the designer? Paley's answer was that we look to the Christian God—the uncaused cause or the undesigned designer. Kant was a philosopher and not a theologian. Hence he needed to smuggle this concept of God in through the back door, so to speak. Thus for Kant, when we seek to understand works of nature *as fully as possible* we must look "beyond the blind mechanism of nature and refer it [the individual work of nature] to a supersensible basis as determining it."[6] The "supersensible basis" Kant is speaking of is a realm completely beyond the boundaries of nature itself. It is a realm beyond our five senses, however much they may be augmented by radio telescopes, electron microscopes, or any other instruments. In short, it is the heavens without a personal God.

Richard Dawkins will have none of this. His book is written for the sole purpose of advancing the argument that, even though works of nature may look like they have been designed for some purpose, and even though we may not be able fully to understand them unless we approach them *as if* they were designed for some purpose, the truth is that the "blind mechanism of nature" that Kant refers to is enough to explain the existence of literally everything in the universe, including life itself. But perhaps that statement needs to be modified slightly. Dawkins concedes that for life to be possible this blind mechanism of nature must be "deployed in a very special way."[7] It must work as the process of *natural selection*. Still, natural selection is not Paley's God or Kant's supersensible basis. It is simply the forces of nature deployed in that "very special way." Thus Dawkins writes, "Natural selection, the blind, unconscious, automatic process which Darwin discovered, and which we now know is the explanation for the existence and apparently purposeful form of all life, has no purpose in mind. It has no mind and no mind's eye. It does not plan for the future. It has no vision, no foresight, no sight at all. If it can be said to play the role of watchmaker in nature, it is the *blind* watchmaker."[8]

That explains the title of Dawkins's book and the first half of the title of this part of my book. Today, despite books like *The Blind Watchmaker*, the truth is that the American public is at best ambivalent about the theory of evolution. But, notwithstanding this ambivalence, in order to understand evolutionary psychology it is necessary at least to *understand* the process of evolution through natural selection. Thus I want now to provide a very brief primer on the subject.

To put it in the most basic language possible, for evolution through natural selection to occur three conditions must be present: random variation, replication, and selection pressure. Notice that the term *gene* does not appear on this list. We will see why shortly. For now, I want to distinguish random variation from pure randomness. To say that evolution through natural selection required *pure* randomness—that is, random action that could not be accounted for by any known or conceivable force of nature—would be to smuggle a concept of God, or at least of "unnaturalness," back into the theory of evolution. To put this point another way, nothing in the theory of evolution requires that any laws of physics be violated at any time. Nothing requires that atoms *not* attract other atoms as they just naturally do under various physical laws, or that molecules *not* bind with other molecules as they just naturally do under various physical laws. The random variation referred to above is "random" only in the sense that it has no knowledge of the selection pressures acting in any environment. It is random with respect to the *consequences* of its variation, and hence varies without regard to the environment. Using the language of statistical correlation discussed earlier, one could say that the individual variations *as such* that occur in each generation show absolutely no correlation with any aspect of selection pressure within the given environment. Thus the randomness of random variation is similar to the randomness one encounters at the roulette table. In both cases, an extreme sensitivity to initial conditions, coupled with the practical impossibility of stating all the influences on a given variable, means that the occurrence of any particular variation, or the likelihood that the ball will drop in any one of the thirty-eight slots, is *essentially* random.

The nonoccurrence of the term *gene* in the above description of evolution draws our attention to the fact that it is not genes as such that are important to evolution. Any "thing" that can reproduce itself—that can *replicate*—can evolve if the conditions are right. One of the conditions needed for evolution (not just replication) to occur is, as I said, random variation. If we now put these two conditions together we see that for evolution to occur the "replicator" must at least show some probability of varying, however slightly, with each replication. A perfect replicator could never *vary*, and hence could never evolve. Evolution thus requires some degree of change, or mutation, in the process of replication.

There is one final point that needs to be understood before one can fully appreciate the theory of evolution. The point is essentially mathematical. Suppose you take one replicator, and it replicates. The result is that you now have two replicators. If each of those two replicators replicates, you have four replicators. If each of those four replicators replicates, you have eight replicators. The reader can see where this is heading. To illustrate this, there is a story, surely apocryphal, but simply too entertaining not to mention.

It is said that the inventor of the game of chess, whoever that may have been in the distant past, was one day brought before the sovereign of the land. The sovereign was pleased with the game and asked what gift the inventor might want as a gesture of appreciation for having invented such a wonderful diversion. The sovereign was prepared to be quite generous. There was gold and silver aplenty in the land. The inventor, however, was humble, saying that nothing much was desired—a mere trifle, really. The inventor, it seemed, would be quite delighted if the sovereign would merely provide from the royal storehouse one grain of wheat to be placed on the first square on the chessboard, two grains to be placed on the second square, four grains to be placed on the third square, eight grains on the fourth square, and so on, until the sixty-fourth square was covered with the required amount of wheat. The sovereign laughed out loud, wondering how the inventor of such a clever game could foolishly want so little in return. But you can guess the ending of the story. To fill just the last square on the chessboard would have required *roughly* eighteen thousand trillion grains of wheat. It is not known how many squares were actually filled before the sovereign caught on and had the inventor fed to the lions.

This story nicely illustrates the geometric increase that populations of replicators can undergo. Even if we assume a very slow rate of replication, the law of geometric progression tells us that in a fairly short time even slow replicators would reach the resource limits of any given environment. Now suppose we call a *selection pressure* anything that prevents the replication of some replicators in some environment. Notice that replicators themselves might well function as selection pressures on other replicators. To survive and replicate in such an environment a given replicator may need to change, or evolve, in some complex ways. Over time, all this can begin to look strangely like intelligence and design.

Something like the idea I have just sketched of evolution through natural selection was worked out independently in the middle of the nineteenth century by Alfred Wallace and Charles Darwin. I feel obliged at least to mention Wallace's name because it is so rarely connected with the theory of evolution these days. For a number of reasons, including possibly a high degree of social intelligence, and perhaps the right social connections, Darwin's name is the one that is today almost exclusively associated with the initial development of the theory of evolution. But, despite what some (lesser minds?) might like to believe, it was not mere political cleverness or social connections that caused Darwin's name to become synonymous with the development of the theory of evolution. The principal reason for this connection in the minds of most individuals must surely be Darwin's prodigious production of popular-science books including, especially, *The Origin of Species*. We would do well to remember that *The Origin of Species* really did sell out on the very first day of its publication in 1859. The book itself ran through six editions, from 1859 to 1872. These editions show the evolution of Darwin's own thinking about evolution.

In addition to these editions of *The Origin of Species*, Darwin also produced a massive (nine-hundred-page) work entitled *The Descent of Man, and Selection in Relation to Sex*, which was published in 1871. The contemporary evolutionary psychologist Geoffrey Miller has called this work Darwin's "crowning achievement."[9] And only one year after this achievement Darwin published another major work entitled *The Expression of Emotions in Man and Animals*, which Christopher Badcock, in his 2000 book *Evolutionary Psychology: A Critical Introduction*, has called "Darwin's pre-eminent work on evolutionary psychology."[10]

These works represent only a small portion of the popular-science writing on evolution that Darwin produced between the publication of the first edition of *The Origin of Species* and his death in 1882. Although it may not be possible to judge a book by its cover, it is possible to judge an author by his influence. Using this criterion, Darwin was surely one of the most influential thinkers ever. How influential? In 1999, the editorial board of the A&E television network (the folks who produce the popular *Biography* series) drew up a list of the top one hundred individuals alive during the second millennium who had "done the most to shape our world today."[11] Those in the book publishing business can take pride in the fact that Johann Gutenberg was first on the list. But Darwin was fourth, behind Isaac Newton and Martin Luther. Even so, in his 1999 book *Darwin's Ghost*, the British geneticist Steve Jones asserted that "*The Origin of Species* is, without doubt, the book of the millennium."[12]

It cannot be emphasized too strongly that the overwhelming majority of books Darwin published—with the possible exception of works like his two-volume "monograph" on barnacles—were written for the educated, but non-scientific, public of his day. In addressing this general audience Darwin quite clearly attempted to construct persuasive scientific arguments, rather than simply recount facts or present dry theories. As a number of commentators have pointed out, Darwin begins *The Origin of Species*—a work that he knew would meet with intellectual "resistance" among his readers—not with a discussion of natural selection, but rather with a discussion of "Variation under Domestication" and "Selection by Man."[13] Darwin wants his readers to *vividly* appreciate what they already know: that every variety of "domestic" dog, or horse, or even pigeon that exists today is a "work" of human animal breeders. Although Darwin does not use this exact example or language, it is truly a marvel that Chihuahuas and Great Danes share the same common ancestor of wolves, and that the existence of these two dog breeds is solely the result of thousands of years of human intervention by breeders who at first carefully noted the minor variations among individual animals, and who then magnified the variations they desired through selective breeding, with the result being the various breeds we have today.

Only when Darwin feels that his audience has sufficiently appreciated the power of *artificial* selection does he move to a discussion of *natural* selection. When he arrives there he is apparently confident enough in his theory to flout the following advice on persuasion given by Aristotle in his *Rhetoric*: when constructing an argument, "we should not . . . put the conclusion itself as a question, unless the balance of truth is unmistakably in our favour."[14] But Darwin asks, "Can the principle of selection, which we have seen is so potent in the hands of man, apply under nature?" He then states the main idea of his theory in the following questions:

> Can it, then, be thought improbable, seeing that variations useful to man have undoubtedly occurred, that other variations useful in some way to each being in the great and complex battle of life, should occur in the course of many successive generations.[*sic*] If such do occur, can we doubt (remembering that many more individuals are born than can possibly survive) that individuals having any advantage, however slight, over others, would have the best chance of surviving and of procreating their kind? On the other hand, we may feel sure that any variation in the least degree injurious would be rigidly destroyed. This preservation of favourable individual differences and variations, and the destruction of those which are injurious, I have called Natural Selection, or the Survival of the Fittest.[15]

I should note here that I am quoting from the Random House Modern Library edition of *The Origin of Species*, which reproduces the sixth edition of that work. In the first edition we do not find the phrase "survival of the fittest."

The phrase itself was not Darwin's but rather that of the nineteenth-century social philosopher Herbert Spencer. There is no doubt that Spencer was an enthusiastic promoter of Darwin's ideas. What Darwin thought of Spencer may have been more complicated. Although Spencer was highly influential during his day, it is fair to say that his reputation has fallen on hard times. Doubtless this has much to do with the fact that Spencer's strenuous intellectual efforts to apply the idea of "survival of the fittest" to the economic and social realms may now seem like merely transparent attempts to justify greed and selfishness by insisting that they are somehow natural. But, that said, nature itself can often seem rather more cunning and cruel than cute and cuddly. Or rather, what seems cute and cuddly in nature may turn out to be otherwise. A fascinating example of this is provided by Darwin in his discussion of the European cuckoo bird.[16]

European cuckoos are parasitic birds. Each cuckoo lays one egg in the nest of another bird—a bird of a different species—and then leaves. The egg hatches and then the baby cuckoo uses the color and shape of its gape as a sort of visual "drug" to induce the mother bird in whose nest it has hatched to feed it. One might think that the baby cuckoo would be content to exploit its foster-parent in this way without causing any more mischief. But baby cuckoos do not want any competition from their foster-siblings. Unfortunately for these foster-siblings, the cuckoo has one big advantage over them. It tends to hatch faster than the other birds. So there is the baby cuckoo, freshly hatched, still blind, tiny, weak, and nearly helpless. What is the *first* thing this little bird does? Instinctively, it moves toward the first egg it can sense. It then manages to roll the egg onto the hollow of its back, taking care to balance the egg with its wings. With a heavy egg on its back, the tiny little bird then backs up to the edge of the nest and ejects the egg over the side. This process is methodically repeated until every egg has been ejected from the nest. Now *that* is survival of the fittest. Frankly, that little performance sounds quite a bit more troubling than the working of any economic or social arrangement Spencer ever defended. Darwin himself seemed not to have liked what he saw in the behavior of baby cuckoos. But he insists nonetheless that this is the way natural selection works. Thus he writes,

> With respect to the means by which this strange and odious instinct was acquired, if it were of great importance for the young cuckoo, as is probably the case, to receive as much food as possible soon after birth, I can see no

special difficulty in its having gradually acquired, during successive generations, the blind desire, the strength, and structure necessary for the work of
ejection; for those young cuckoos which had such habits and structure best
developed would be the most securely reared.[17]

My point here is not to get into a philosophical discussion of the ethical
nature of nature. Darwin well understood that his theory might pose some
problems for those, like Paley, who believed that God's benevolence was everywhere on display in the world around us. But Darwin simply does not discuss
these problems in *The Origin of Species*. It is not that kind of book—although,
as we shall see at the end of this chapter, Darwin could sometimes wax poetic
about nature.

We will come to Darwin's poetry later. But for now, I want to turn to a
more detailed discussion of evolution through natural selection. The following discussion will pay particularly close attention to some aspects regarding
the mechanisms and the speed of evolution. As a way into this discussion, I
want to begin by considering one of the most exquisite examples of the sheer
power of evolution and one of the most awe-inspiring organs of the human
body: the eye. With the exception of the brain itself, no other organ of the
human body elicits as much fascination, on so many levels, as does the eye. In
fact, one could justifiably claim that the eye is really part of the brain, since the
retina is actually brain tissue that moves to its place in the back of the eye
during fetal development.

Putting this interesting physiological point aside, however, there is no question that over millions of years animals that had the ability to see had a clear
survival advantage relative to those that did not. Human eyes in particular are
extremely well adapted to enable us to differentiate shapes and edges, to
perceive depth and thereby judge distance, to discern colors in various lighting, and to process visual information in thousands of other ways. All of these
visual abilities would have been advantageous in innumerable ways to ancestral humans. Of course, there are limits. Human eyes are not equipped with
a built-in "bug detector" as are frog eyes, or with a built-in "hawk detector" as
are rabbit eyes.

It is easy to see (no pun intended) why the human eye does not come
equipped with a bug or hawk detector. Neither of these animals would have
been *especially* important to us *in the ancestral environment*. During the vast
majority of the time the human eye was evolving, humans were subjected to

different selection pressures than the selection pressures that faced frogs or rabbits. This helps to illustrate a point that is critical to evolutionary psychology in general. It is a point that I have already repeated in various forms in this book, and that I will continue to repeat in different forms—with the indulgence of the reader. The point is that both the human body and mind are adapted *to the ancestral environment,* not to the environment many (or most) humans face in the twenty-first century. Evolution is a long and slow process. Cultural and technological change, on the other hand, happens exceedingly rapidly in "evolutionary" time. Thus it is well to remember that the selection pressures faced by our ancestors during the overwhelming period of hominoid evolution are not the selection pressures we face today. This point often becomes controversial when applied to a discussion of the human mind. But to see this point in a fascinating, and relatively uncontroversial light, consider the human ability to perceive colors.

In order for this ability to develop through the process of natural selection, it would have needed to confer some survival or reproductive advantage on those who possessed it in the ancestral environment. It undoubtedly conferred many advantages. To find one you need look no further than your local fruit stand. Surely an ability to distinguish the color of fruits—from cherry red to lime green—would have conferred a survival advantage on ancestral humans. Especially in the ancestral environment, such an ability would be extremely useful in identifying a particular fruit and determining its ripeness.

Now here is the key point. Our ability to discern colors developed in an ancestral environment *in which about the only source of light was sunlight.* (Moonlight is, of course, sunlight reflected by the moon.) Under this type of natural lighting, the human eye is remarkably good at discerning many different shades of color and at maintaining the constancy of these colors. This second ability—color constancy—should be amazing. But it seems so "natural" that it is difficult to be amazed. Suppose, however, that you drive up to an intersection at midnight when there is a full moon in the sky. You see that the lights on the traffic signal in the distance turn from green, to yellow, to red. These are all colors that you can easily discern in the moonlight. But now you go back to that same intersection twelve hours later when the sun is shining brightly and there is not a cloud in the sky. The light reaching your eyes from that traffic signal is now a complex mixture of the light from the signal and the environment. This mixture is vastly different from the one you experienced twelve hours earlier. But you still see the lights as green, yellow, and red. The reason you are able to maintain this ability to see the same color constantly under different lighting conditions is that the mind makes assumptions about the type of lighting conditions that can exist in the world—that is, the world of our ancestors long before there was artificial lighting. As it happens, under *most* artificial lighting color constancy can still be maintained. But many parking lots and indoor garages are illuminated by sodium or mercury vapor lights that radiate a narrow range of wavelengths of light, the intensity of which could never have existed in the ancestral environment. Hence, under these lights the human eye loses much of its ability to discern color. The result can be that both your cherry red Honda and the lime green Toyota next to it look as though they are the same shade of yellowish green.[18]

The point of this example is to show how selection pressures can drive the design of organs, like the eye, that evolved over the course of millions of years. Of course, we do not know precisely how the eye evolved. But evolution through natural selection tells us that it must have evolved in a step-by-step process, as random variations in, for example, genes that enable color vision, or genes that enable stereoscopic vision, were selected *for* in a given environment because they increased (however minutely) their bearer's chances—and hence, ultimately, their chances—of reproduction in the given environment. Thus the evolutionary psychologists John Tooby and Leda Cosmides insist that "it is easy to see how selection, through retaining those accidental modifications that improved performance, could start with an initial accidentally light-sensitive nerve ending or regulatory cell and transform it, through a large enough succession of increasingly complex functional forms, into the superlatively crafted modern eye."[19]

Perhaps it is not *quite* as easy to see this as Tooby and Cosmides assert. But their broader point is unimpeachable. Evolution through natural selection is the only scientific way of explaining the structure and functioning of the human eye. Does this mean that the human eye is "superlatively crafted," as Tooby and Cosmides claim? I am tempted to say that the beauty of the eye's craftsmanship is decidedly in the eye of the beholder. I am also tempted to say that Tooby and Cosmides may have something of a blind spot when it comes to evaluating the design of the eye.

I realize these puns are awful and unnecessary. But the point I now want to make is important. Since evolution is a step-by-step process, driven by random variations at each step, and then culled by selection pressure, also at each step, the organisms and parts of organisms that evolution "designs" need not—in fact, almost certainly will not—show the kind of forethought that we see in the design of human artifacts, like watches or airplanes. This is both good and bad. It is good because, in lacking any forethought, or any thought at all, evolution cannot become trapped in a particular thought pattern. Evolution cannot get "hung up" on one type of design, or one type of solution to a given survival problem. It does not need to be told that it must "think outside of the box," because it is not in a box in the first place.

On the other hand, the fact that complex designs, like the eye, are built, step by step, means that each individual step in the design process must confer an immediate survival or reproductive benefit on the organism. There can be no waiting for several steps to converge on one really good design. This is obvious for two reasons. First, evolution has no foresight. So it could never "plan ahead," even if such planning could produce a really good design. But, second, even *if* evolution could "plan ahead," it would still need to follow the above rule that each step in the process must confer a survival or reproductive advantage on the organism. Suppose that evolution were able to plan ahead, in the sense of planning a fifty-step process that would result in a given organism's possessing a better eye than it had at the beginning of the process. Now suppose that at the forty-ninth step in the process a slight survival *disadvantage* were conferred on the organism. It would make no difference if that disadvantage could be remedied by the fiftieth step, because the organism could, by definition, never get to that step. Thus, to repeat, no step in the process could confer a survival or reproductive disadvantage on the organism.

But what about a step that conferred neither a survival advantage nor disadvantage: a neutral step. Say evolution wanted to produce a nice, clear, well-formed lens and covering for the light-sensitive cells of the eye, before it actually produced those cells. In the long run this might be advantageous to the organism in that it would protect those light-sensitive cells when they did develop, and hence might allow them to develop that much faster. Is not this an example of a neutral step that evolution could take? The short answer is no. Anything that is part of an organism draws on the metabolic energy of the organism. But something that does not increase the survival or reproductive success of the organism, yet eats metabolic energy nonetheless, is at that moment, by definition, wasteful. And waste is a disadvantage.

Because evolution must take steps in the manner described above, organisms may evolve organs that can appear rather poorly designed in retrospect.

Again, consider the evolution of the human eye. Start by imaging a small creature crawling around a few million years ago. This poor slob of a life form—the ancestor of you and me—gets along tolerably well from day to day without anything like the ability to see. Then a random variation in one of its genes produces in it a small cell, located below its skin, that is sensitive to gross changes in light—from, say, the brightness of noon to the darkness of midnight. For any number of reasons, this variation gives the creature a *minute* survival and reproductive advantage over its fellows.

Now suppose that the nerve ending leading from this light-sensitive cell happened to be positioned on top of the cell, over the very surface that detected the light. This would be a design flaw, to be sure. *But it would still represent an improvement on total blindness.* Hence, it would still confer some slight advantage on its bearer, and it could therefore evolve. Suppose further that after a series of random variations this organism develops a translucent coating over that light-sensitive cell, and then a lens to help focus the light on that cell, and then muscles around the lens which give the organism the ability to change the focal length of the lens, and so forth, over the course of millions of years. Eventually, you would end up with the human eye. But, the eye would possess that little design flaw—*repeated roughly three million times.* What do I mean by this? Let me use an analogy Richard Dawkins develops in *The Blind Watchmaker*.

Think of the eye as a simple machine. Imagine the retina of this machine eye—the part on which the light is focused—as a curved metal surface containing millions of light-sensitive "photocells." Now imagine that a wire runs from each one of these photocells into a big cable, and then to a computer that processes the information. Here is a question. If you were designing this machine, on which side of the photocell would you attach the wire that runs from it to the computer? Would you attach the wire on the side that faces toward the light—that is, in front of the photocell—or on the side that faces away from the light—that is, behind the photocell? If you answered that you would put the wire behind the photocell, you may be "smarter" than evolution. Astonishingly, the "wires" that are attached to the cells of the retina of the human eye run in *front* of the retina. They all converge on a single spot and then punch a hole in the retina, forming a "cable," called the optic nerve, that leads to the brain. Not only do the "wires" obviously interfere with the collection of light on the retina, but the hole through which they must exit creates a blind spot. No engineer would ever design a "seeing machine" in this way. Is this evidence of bad design? Of course not. It is evidence of design by natural selection.[20]

If you still cannot bring yourself to believe that evolution through natural selection alone could produce something as intricate as the eye, do not be distressed. Darwin confesses in *The Origin of Species* that even he had difficulty believing this. But, according to Darwin, if you just let reason be your guide, you will see, as you must, the length to which evolution can reach in explaining our world. Thus Darwin writes, "To arrive . . . at a just conclusion regarding the formation of the eye, with all its marvelous *yet not absolutely perfect* character, it is indispensable that the reason should conquer the imagination; but I have felt the difficulty far too keenly to be surprised at others hesitating to extend the principle of natural selection to so startling a length."[21]

So far we have seen how the mechanism of evolution through natural selection, responding to selection pressures in a given environment, works through a step-by-step process to fashion a solution or solutions to those selection pressures. I have stressed repeatedly that the process upon which this is based involves random variation. But this is a difficult point to grasp exactly, and sometimes the mind slips. There is undoubtedly a clever example involving the evolution of the human eye which could serve to illustrate the slippage I am talking about. But since I cannot think of it, I will fall back on the standard example.

Imagine the ancestor of a giraffe, some millions of years ago. Suppose this animal looked pretty much like a horse and went about eating the foliage from trees. One can well imagine that an ancestral giraffe who stretched its neck every day in an attempt to get more and better food might thereby elongate its neck. One might also imagine that this ancestral giraffe, with its now slightly elongated neck, might pass this trait on to its descendants. If these descendants behaved as their forebears had, stretching their necks to obtain more or better food, it seems clear how modern-day giraffes obtained their long necks. One could well imagine that evolution works in such a way that organisms consciously desire to obtain some goal, use their bodies to obtain the goal—sometimes thereby altering their bodies—and then pass whatever alterations they undergo on to their descendants. This theory has the virtue of making evolution "reward" use and punish disuse.

Some readers may remember being taught that something like this explanation for the working of evolution was advanced at the end of the eighteenth century by Jean-Baptiste Lamarck. Actually, Lamarck's now famous connection to the giraffe's neck may be much exaggerated, as Stephen Jay Gould pointed out in a wonderful essay entitled "The Tallest Tale."[22] But, as Dawkins notes, Lamarck should be remembered for at least advancing *some* theory of evolution. (Remember, neither Paley, nor Kant, nor any number of other intellectuals at the time could manage even this.) Unfortunately for Lamarck, he is remembered primarily for advancing a fallacious theory of evolution. Put simply, the Lamarckian fallacy involves the assumption that acquired traits can be inherited. They cannot. But, as I said, this is at times a difficult idea to grasp firmly. (How many times have you heard someone note that the human index finger is getting longer from one generation to the next *because* we use that finger so much to punch buttons on telephones and computer keypads?)

The fact that acquired traits are not inherited is fortunate, however, for there are almost certainly many more ways that acquired traits can pose disadvantages, rather than advantages, to a body. Thus a running back who destroys his knees in the process of winning several Super Bowls, and who therefore walks with a limp at the age of thirty, can be sure that he will not pass that limp on to his children. There is, it turns out, something of a "firewall" between the "germ cells" of the body (the gonads and ovaries, which produce cells that develop into sperm and eggs in males and females, respectively) and the rest of the body. With but few exceptions—radiation poisoning being an important one—nothing that happens to the body can affect these germ cells *as such*. Of course, the *propagation* of these cells can be affected by what happens to the body if, for example, the body happens to die before it can procreate.

So Lamarck was wrong. What one does not use one's descendants do not lose. If this were not the case, then Jewish males born today would lack foreskins. But this proves only that Lamarck was wrong about the *mechanism* of evolution. Giraffes, after all, do have long necks. But they do not have these necks *because* their ancestors stretched for food. They have them because, through random variation and selection pressure, those ancestral giraffes with long necks left more descendants than their fellows with shorter necks. Random variation and selection pressure drive evolution, not the inheritance of acquired traits.

With that said, I want to turn from a discussion of the mechanisms through which evolution operates to a discussion of some factors that determine the speed of evolution itself. One factor that can determine the speed of evolution is the rate of reproduction and random variation—that is, mutation—of a given organism. Thus, as Steve Jones points out, evolutionary theory explains perfectly the reason that the AIDS epidemic is so difficult to control. The virus that causes AIDS reproduces and mutates very rapidly, thus allowing it to stay one step—probably several steps—ahead of our attempts to destroy it. Simply put, the AIDS virus evolves rapidly to meet the demands of the various selection pressures arrayed against it. Those pressures include the body's own defenses and the various drugs we use to combat the virus.[23]

Another factor that can affect the speed of evolution is the rate at which selection pressures change in a given environment. If, in a given organism, random variation (mutation) is slow and selection pressures change quickly, that organism can be in serious trouble. Dinosaurs are fossils today apparently because they could not adapt quickly enough to the rapid change in selection pressures that occurred in their environment—probably due to a meteor impact that kicked up dust, cooled the planet, and caused plant life to die, thus starving dinosaurs into extinction.

The mathematical relationships between the rate of change of random variation and the rate of change of selection pressures can become amazingly complex and can lead to some fascinating implications for the speed of evolution, as we can see by examining one type of selection pressure: sexual selection. Although this topic gets only a very brief mention in *The Origin of Species*, Darwin follows that discussion with a roughly six-hundred-page analysis of the subject in *The Descent of Man, and Selection in Relation to Sex*. As I noted earlier, the evolutionary psychologist Geoffrey Miller takes this later work as Darwin's crowning achievement, precisely because most of it deals with sexual selection. Indeed, Miller makes sexual selection the focus of his 2000 book *The Mating Mind: How Sexual Choice Shaped the Evolution of Human Nature*. With wit and style, Miller advances the audacious argument that sexual selection in humans can account, better than any other explanation offered by rival evolutionary psychologists, for all of the important aspects of human beings: our general creativity, our ability to use language, our ability and desire to create art, our various moral and ethical systems, and so forth.

In chapter seven I will discuss more fully Miller's particular theory. But for now I want to examine some of the fascinating implications of sexual selection for the speed of evolution. To do so, suppose that tomorrow, because of some random variation, a sufficiently large number of women in a given population develop an innate preference for men with big feet. These women find such men highly attractive and indeed will only mate with men with big feet. (The reason I am stipulating that women possess this trait is important. As we shall see, for a variety of evolutionarily determined reasons, women are the choosier sex.) Obviously, there is now a selection pressure on men. Those with big feet reproduce more than those without big feet. Thus men with big feet will come to predominate in this population.

I need to emphasize here that the preference of these women for big feet in men is a preference that the women are born with, not one they acquire during their lives. Of course, it is conceivable that a preference for men with big feet that was *acquired* by a sufficiently large number of women in a population at some point during their lifetimes and that was *rigorously* passed on by society to the daughters of these women *could* act as a selection pressure on men, thereby *eventually* causing men with big feet to predominate in the population. One could even imagine a law that made it illegal for women to mate with men who did not have big feet. This is, of course, the whole idea behind eugenics. But since the daughters of these women would not be born with this preference for big feet in men, it is not clear how long it could be maintained in the population, even using extreme measures. Adolescents do, after all, delight in rebelling against the fashions of their parents and, perhaps especially, "the system." And the good news about totalitarian regimes is that, while they may be able to control many aspects of people's lives, they cannot (at least for the time being) change human nature.

But for our present purposes this does not matter, since in my example I clearly stated that the preference in question was innate. To be more precise, let us assume it is innate *and highly heritable*. Let us also assume that foot size is highly heritable in males. On the basis of these assumptions we can conclude, as I did above, that men with big feet will come to predominate in the given population precisely because an important selection pressure would favor men with big feet.

But now notice something even more fascinating. Not only will men with big feet become predominant in the population. Women with an innate preference for men with big feet will also become predominant in the population. The reason for this can be seen by examining the fate of the male and female descendants of a woman who happened to have an innate attraction to men with small feet. Such a woman would mate with the few men around who had small feet. Her male children would thus have small feet, and consequently not be able to find mates, thus removing their small-feet genes from the gene pool. But her female children, those with a preference for men with small feet, would also have trouble finding a mate, since the *trend* in the population is now toward the birth of male children who will have big feet. Hence, the female children who did innately prefer men with small feet would also leave fewer offspring, of either sex. On the other hand, the female children of females who preferred men with big feet would have considerably less trouble finding a mate, and hence their preference for males with big feet would get passed on to their female children, just as their mates' genes for producing big feet would get passed on to the male children. What I am describing here is known as a *positive feedback loop*. The attraction by women for men with big feet produces more men with big feet, which reinforces the attraction by women for men with big feet, and so on. This can create what is also called *runaway selection*. Miller discusses this phenomenon at some length. He takes as his example not feet but a different, and uniquely male, organ of the human body. All organs, it turns out, are subject to selection pressure. Some may be more subject than others to sexual selection by women. Size and shape may matter after all.[24]

But back to feet. Obviously, having big feet *could* be advantageous to the survival of a male, irrespective of the increase it might provide to his chances of getting a mate in the environment described above. A male with big feet *might* be able to walk better or land more sure-footedly after a jump. But even if there were some survival advantage to having big feet, surely this survival advantage would exist only up to a point. Beyond that point, having big feet might confer a decided survival disadvantage on males. Now here is a problem: Females might continue to prefer men with big feet. Indeed, they might continue to prefer men with big feet *just because* big feet are a handicap. Thus men would continue to evolve bigger feet, even though such feet were physically disadvantageous to them. This is called the *handicap principle*. To examine it more fully, I will leave our discussion of male feet, and turn to the example most often used in the literature.

The example is that of the peacock's tail. Peacocks receive absolutely no practical benefit (that is, no *survival* benefit) from large tails. These tails do not enable them to flee from predators more effectively, or to hide more effectively, or to use metabolic energy more efficiently. In fact, exactly the reverse is true on all of these counts. A peacock's tail is, therefore, a significant handicap to the peacock. The bigger the tail, the bigger the handicap. So why do peacocks have big tails? Because peahens love them. Peahens know that a peacock who can manage to survive with an outrageously big tail must be quite fit indeed. Thus even though a peacock's tail confers on it a survival disadvantage, it also confers on it a reproductive advantage. And it confers this reproductive advantage just because it confers the survival disadvantage.

Such are the fascinations of sexual selection. Of course all types of selection —both natural and sexual—must produce winners and losers. That is simply the way of evolution. Earlier in this chapter I suggested that Darwin's theory of evolution may not be for the faint of heart. Nature *is* often "red in tooth and claw." The particular behavior of baby cuckoos is only one example of survival of the fittest. There are other examples, perhaps even more troubling. Male langur monkeys, for instance, often kill the infant and young monkeys of troops they have taken over.[25] Darwin was surely as disturbed as anyone by what he saw in nature. All the same, he chose to end his best-known work on what reads like a decidedly optimistic note about life. As an appropriate conclusion to this chapter, I cannot resist quoting the last two sentences of *The Origin of Species*. The optimism and poetry of these two sentences is obvious. Less obvious, but equally interesting, is the contrast Darwin appears to draw between the mechanical regularity of physics and the wonderful unpredictability of biology.[26]

> Thus, from the war of nature, from famine and death, the most exalted object which we are capable of conceiving, namely, the production of the higher animals, directly follows. There is a grandeur in this view of life, with its several powers, having been originally breathed by the Creator into a few forms or into one; and that, whilst this planet has gone cycling on according to the fixed law of gravity, from so simple a beginning endless forms most beautiful and most wonderful have been, and are being evolved.[27]

CHAPTER SIX

What Is the Mind?

Richard Dawkins's general argument in *The Blind Watchmaker* is that evolution through natural selection can account for all life on the planet earth, from tulips to (now extinct) pterodactyls. Additionally, Dawkins argues that natural selection is the only *scientific* way of accounting for all of the various complex organs with which these life-forms come equipped. Eyes are an example of an extremely complex organ possessed in several varieties by many species. Dawkins also mentions the Tadarida bat's marvelously complex ear, which is capable of switching "off" so that the bat does not go deaf when it emits the very loud shrieks it sends out to navigate, then switching back "on" so that the bat can hear the much softer echoes of those shrieks upon their almost instantaneous return, and performing this switching up to fifty times *per second*.[1] Surely Charles Darwin was right; there is a grandeur in this view of life. Much of that grandeur arises out of the astounding complexity that evolution through natural selection is able to achieve.

But there is complexity, and there is complexity. I suspect that many individuals—perhaps many so-called intellectuals especially—do embrace the general theory of evolution, *but only up to a point*. They want to insist that there is one thing that evolution cannot explain. That thing is the human mind. To be sure, evolution may be able to explain the physical structures of the human brain. But the *mind* is different. The special status of the human mind as the one thing that resists explanation by evolution through natural selection can be accounted for in a number of ways, depending upon who is doing the accounting. For some, the mind's special status inheres in its unique capacity for self-awareness or self-reflection. This capacity, it is argued, is not really a physical property of the brain but rather a quasi-metaphysical, perhaps transcendental, quality of the human being *as such*.[2] For others, the mind's special status inheres in its unique capacity for creative thought—particularly as such thought is manifested in all forms of artistic expression. Again, the idea here is that *true* creativity is not a physical property of the brain. It is instead a kind of gift from somewhere outside the physical realm.[3] Obviously, both of these accounts of the mind's special status are compatible one with another and perhaps mutually reinforcing. They both maintain that the *qualitative* complexity of the mind is such that it can never be fully explained even by the best evolutionary theory. This argument is given a kind of final philosophical "lift" with the observation that the best evolutionary theory that could ever be produced would, of course, need to be produced by the very type of mind that the theory itself was trying to explain. But, as a wit once said, if the mind were simple enough to be understood, we would be too simple to understand it.

Steven Pinker and other evolutionary psychologists will have none of this. They flatly reject the idea that the mind itself is inexplicable in evolutionary terms. Indeed, they insist not just that the brain but the mind as well is in the final analysis just another organ of the human body—albeit a very complex one. Still, like any other organ of the body, the mind must have evolved right along with the rest of our material substance. To be sure, the mind's complexity, and maybe its capacity for self-reflection as well, guarantees that it will be a *uniquely* fascinating organ. This explains, Pinker modestly notes, why you are not likely to read a popular-science book with a title like *How the Pancreas Works*. Nonetheless, according to Pinker and other evolutionary psychologists, the mind is best conceptualized as an organ. And like all organs it has a function. *Its function is information processing and computation.* This makes the evolved mind sound very much like an evolved computer. Although Pinker says the computer is not a good metaphor for the mind, he cannot help deploying exactly this metaphor when the allusion is just too good to pass up.[4] Thus, in the beginning of *How the Mind Works* he writes, "Richard Dawkins called natural selection the Blind Watchmaker; in the case of the mind, we can call it the Blind Programmer. Our mental programs work as well as they do because they were shaped by selection to allow our ancestors to master rocks, tools, plants, animals, and each other, ultimately in the service of survival and reproduction."[5]

With everyone playing off of the allusions of everyone else, the point might get somewhat obscured. Hence it is important to remember that when Pinker speaks of our mental programs being shaped by selection he means that they were shaped by the same blind mechanical forces that Dawkins argued shaped all aspects of all life. To be sure, in order to shape our mental programs, these blind mechanical forces had to be deployed in that "very special way" that is evolution through natural selection. But—and this is the key point—it still follows that evolution through natural selection should be able to explain the mind *as well as* it can explain the body. My title for the present part of this book tries to make this key point by combining two metaphors. Granted, I have chosen to modify the figure of speech Pinker uses somewhat—making the programmer scatterbrained rather than blind, since to me this better connotes the randomness upon which evolution through natural selection must draw. But the essential point remains. Evolutionary psychology is the study of the evolved mind, where the mind is understood as an information-processing and computational device.

But what exactly does it mean to say that the mind is an information-processing and computational device? An answer to this question is absolutely critical, for if there is anything that differentiates evolutionary psychology from its sociobiological origins, it is this very specific description of the mind. To get an understanding of what is meant by this description we can begin by turning to what has been called the manifesto of evolutionary psychology: a 1992 volume of essays entitled *The Adapted Mind: Evolutionary Psychology and the Generation of Culture*, edited by Jerome H. Barkow, Leda Cosmides, and John Tooby. In their introduction, the editors begin by noting, "There are various languages within psychology for describing the structure of a psychological mechanism, and many evolutionary psychologists take advantage of the new descriptive precision made possible by cognitive science."[6] They continue by arguing that this new descriptive precision allows evolutionary psychologists

> to describe a brain as a system that processes information—a computer made out of organic compounds rather than silicon chips. The brain takes sensorily derived information from the environment as input, performs complex transformations on that information, and produces either data structures (representations) or behavior as output. Consequently, it . . . can be described in two mutually compatible and complementary ways. A neuroscience description characterizes the ways in which its physical components interact; a cognitive, or information-processing, description characterizes the "programs" that govern its operation. In cognitive psychology, the term *mind* is used to refer to an information-processing description of the functioning of the brain, and not in any colloquial sense.[7]

From this the editors conclude: "An account of the evolution of the mind is an account of how and why the information-processing organization of the nervous system came to have the functional properties that it does. Information-processing language—the language of cognitive psychology—is simply a way of getting specific about what, exactly, a psychological mechanism does."[8]

What the editors of *The Adapted Mind* call a *psychological mechanism* is also referred to among evolutionary psychologists as a *mental module* or *mental organ.* Thus Pinker writes, "The mind is organized into modules or mental organs, each with a specialized design that makes it an expert in one arena of interaction with the world. The modules' basic logic is specified by our genetic program."[9] But having said that, Pinker adds,

> The word "module" brings to mind detachable, snap-in components, and that is misleading. Mental modules are not likely to be visible to the naked eye as circumscribed territories on the surface of the brain, like the flank steak and the rump roast on the supermarket cow display. A mental module probably looks more like roadkill, sprawling messily over the bulges and crevasses of the brain. Or it may be broken into regions that are interconnected by fibers that make the regions act as a unit. The beauty of information processing is the flexibility of its demand for real estate. Just as a corporation's management can be scattered across sites linked by a telecommunications network, or a computer program can be fragmented into different parts of the disk or memory, the circuitry underlying a psychological module might be distributed across the brain in a spatially haphazard manner.[10]

I grant that all this talk of mental modules, information processing, and computation can begin to sound extremely passionless and sterile. What Pinker and his colleagues seem to be offering us is not only an incredibly reductionistic account of the mind but an incredibly rationalistic one as well. To be sure, these are real criticisms of evolutionary psychology. But it must be pointed out that, in addition to being reductionistic and rationalistic, the theory of the mind offered by evolutionary psychologists does have at least one important virtue. It is a *pragmatic* theory. Pinker makes this point repeatedly, in a number of contexts, throughout *How the Mind Works*. Granted, on occasion Pinker does seem to be exaggerating the uselessness of some theories of the mind that might be competitors to his own theory. But, truth be told, there are a number of theories of the mind floating around that seem to be a bit on the fuzzy side. Thus it is understandable that Pinker would write, "The entities now commonly evoked to explain the mind—such as general intelligence, a capacity to form culture, and multipurpose learning strategies—will surely go the way of protoplasm in biology and of earth, air, fire, and water in physics. These entities are so formless, compared to the exacting phenomena they are meant to explain, that they must be granted near-magical powers."[11] Pinker and his colleagues think they can do better than past theorists when it comes to thinking about the mind. One of the strengths of reductionism, after all, is that it at least provides a place to begin when attempting to formulate concepts and devise theories. Again, Pinker writes: "Thoughts and thinking are no longer ghostly enigmas but mechanical processes that can be studied, and the strengths and weaknesses of different theories can be examined and debated."[12]

Again, reductionism is the key to evolutionary psychology's explanatory power. Evolutionary psychology reduces thought to mental representations and to mental operations (i.e., algorithms) that are performed on those representations. In this way, evolutionary psychology can achieve the same level of scientific rigor about its discussions of the mind as biology can achieve about its discussions of the body. Just as biologists today no longer need speak of the four bodily "humors" (blood, phlegm, black bile, and yellow bile), but can instead get very specific about hormones and their functions, so too evolutionary psychologists no longer feel a need to use vague terms like *understanding*, *knowledge*, or *comprehension*. Instead, they claim that they can explain mental processes in much more exact terms—the terms of mental representations and information processing. Thus Pinker writes:

> Pinning down mental representations is the route to rigor in psychology. Many explanations of behavior have an airy-fairy feel to them because they explain psychological phenomena in terms of other, equally mysterious psychological phenomena. Why do people have more trouble with this task than with that one? Because the first is "more difficult." Why do people generalize a fact about one object to another object? Because the objects are "similar." Why do people notice this event but not that one? Because the first event is "more salient." These explanations are scams. Difficulty, similarity, and salience are in the mind of the beholder, which is what we should be trying to explain. A computer finds it more difficult to remember the gist of *Little Red Riding Hood* than to remember a twenty-digit number; you find it more difficult to remember the number than the gist. You find two crumpled balls of newspaper to be similar, even though their shapes are completely different, and find two people's faces to be different, though their shapes are almost the

same. Migrating birds that navigate at night by the stars in the sky find the positions of the constellations at different times of night quite salient; to a typical person, they are barely noticeable.[13]

The key point here is one about rigor. Much of contemporary psychology lacks rigor because it continues to use vague terms to explain psychological phenomena. "But," Pinker insists,

> if we hop down to the level of representations, we find a firmer sort of entity, which can be rigorously counted and matched. If a theory of psychology is any good, it should predict that the representations required by the "difficult" task contain more symbols (count 'em) or trigger a longer chain of demons [i.e., computational operations or algorithms—EMG] than those of the "easy" task. It should predict that the representations of two "similar" things have more shared symbols and fewer nonshared symbols than the representations of "dissimilar" things. The "salient" entities should have different representations from their neighbors; the "nonsalient" entities should have the same ones.[14]

So much for airy-fairy entities and vague terminology. As one might expect, Pinker's ultrareductionistic and very pragmatic view of thinking explains rather well the omission of any abstract philosophical discussion in *How the Mind Works*. At first glance, this omission might seem rather odd. After all, a book about the mind would clearly lend itself to untethered speculations about thought and meaning and consciousness. Similarly, such a book would seem to present a good occasion for the consideration of deep ethical and moral issues. But such speculations and considerations are almost wholly absent from Pinker's book. To be sure, any discussion of the mind will necessarily *imply* views on philosophical matters like the origin of consciousness, or the problem of free will and moral responsibility, or, perhaps especially, the ability of humans to understand our own most fundamental nature. Such a discussion will therefore necessarily embody a moral and political philosophy. Reading between the lines of *How the Mind Works*, it is possible to tease out this philosophy. To put it briefly, that philosophy appears to embody something like the standard, educated-American, individualistic, rights-based, liberal-democratic, secular, moderately egalitarian, and overwhelmingly pragmatic orientation toward life, with a touch of irony and urbane sophistication thrown in for good measure.

But as I said, Pinker is nowhere explicit about this philosophy in his book, and I am sure he would rather leave speculation of the sort in which I have just engaged to his more philosophically oriented critics. True to its title, the focus of *How the Mind Works* is on the working of the mind—this metaphorical computer—and specifically how it can be understood as performing fascinatingly complex feats of information processing and computation. When reading Pinker one gets the sense that one is in the presence of an incredibly smooth and incredibly knowledgeable computer salesman who genuinely believes in the product he is selling and who is determined to persuade you to purchase it by getting you fascinated in all the bells and whistles with which it comes equipped. One does not at all get that sense when reading, for example, Edward O. Wilson's *Consilience*—the other popular book on evolutionary psychology that appeared in the late 1990s. Unlike Pinker, Wilson is like your favorite white-haired old college professor whose lectures you could listen to for hours, probably due to the sheer force of their erudition, but who is also genuinely—even agonizingly—concerned that you take in at an almost spiritual level what is being said.

To get a better sense of how the mind is represented in books like *How the Mind Works* and *The Adapted Mind*, and to make what I have said thus far in this chapter a bit more concrete, I want now to consider briefly one of the most interesting of the mental modules that these books tell us come as "standard equipment" (to use Pinker's phrase) on all human minds. The module I want to examine is used for determining one's optimal investment in one's offspring. Call this a *parental investment module*. It is quite obvious how such a module could have evolved. The genes of parents who were best able to optimize their resource investments in their children would get passed on in the form of those very children. It also seems obvious that this module would be different for mothers and fathers. Mothers would clearly be more likely than fathers to invest heavily in their children. Even beyond the obvious fact that in the ancestral environment only the mother (or another female) could nurse a small child, a mother would realize that any one of her children would represent a much greater fraction of her overall reproductive potential than any one child would represent of a father's overall reproductive potential. The father might also need to discount his investment in any child based on the uncertainty he had about whether the child in question was actually his biological offspring. Where more than one child in a given family were present the situation would become even more complicated. Parents would need to calculate their relative investments based on such factors as their children's age, health, sex, and so forth. A parent who strictly *equalized* all of her daily parental investment among her several children would almost certainly be following a suboptimal investment strategy. A sick child, for example, might temporarily need a substantial increase in parental investment. On the other hand, if resources become extremely scarce, parents might be faced with the ultimate choice. Here Sarah Blaffer Hrdy—who has been called a feminist evolutionary psychologist—draws our attention to the practice of "widespread maternal abandonment of infants" in Europe from late antiquity until roughly the Renaissance. In her book *Mother Nature: A History of Mothers, Infants, and Natural Selection*, Hrdy argues that far from being "abnormal" or "sick" individuals, these women who abandoned their children were responding to extreme circumstances (notably the existence of desperate poverty and disease all around them) with extreme but rational measures. Hrdy's point is that we ought not to condemn these mothers. Rather we should seek to understand that in the face of desperate circumstances they needed to—and did—make some critical calculations about parental investment.[15] The broader point for evolutionary psychology is that even in less desperate circumstances parents make decisions—and often very complex decisions—about how much to invest in their children. Parents, it is argued, come equipped with a mental module that makes it possible for them to calculate these investments.

Unfortunately, I suspect that this description of the working of just one mental module may have turned more than a few readers off to the whole idea of evolutionary psychology. Could Diogenes himself have dreamed up a more cynical view of parenting? The claim that parents think of their children in terms of resources and investments will seem altogether too cold and calculating or at least too myopic to be of much use. But evolutionary psychologists do not claim *exactly* this. Instead they emphasize that mental modules often— indeed almost always—operate at a level below our conscious awareness. To some extent, this may seem to be morally exculpatory insofar as it suggests that we are not always or often aware of the calculations our mental modules are performing. Hence, when considering parental investment strategies, mate selection preferences, or a host of other life options, humans do not *consciously* run programs that churn though a series of algorithms, thereby producing some output. We may be calculating beings, but rarely are we aware of these calculations. Thus it is not as if the average American couple puts their children to bed at night and then sits down at the kitchen table with pen and paper to plan out tomorrow's parental investment schedule. To be sure, parents *do* make conscious decisions about, for example, how much to invest in their child's college fund versus how much to invest in their own 401(k). And these broad decisions are clearly explainable by reference to a parental investment module, even though parents in the ancestral environment never knew of colleges or retirement funds. But just as surely a parent who stays home from work to care for a child with the flu does not consciously think, "On balance, this is a wise parental investment." The parent thinks, "My child is sick."

This discussion of the parental investment module helps to make clear one of the most significant differences between evolutionary psychology and what one might call "conventional" psychology. As a glance at most introductory college textbooks in psychology shows, conventional psychology represents the mind quite differently than does evolutionary psychology. Conventional psychology stresses the study of general attributes of the mind like memory, attention, intelligence, personality, and so forth.[16] The problem with this approach is that it fails to explain the purpose of these attributes. Of course, one could insist that they have no purpose, nor does the mind with which they are connected. Another way of putting this point is to say that the mind is an *all*-purpose "organ" that can adapt to any environment and pursue any goal or goals it chooses. At first glance, this view of the mind can seem quite appealing. It implies that minds (and hence the humans who possess them) are radically free to choose our own paths. The opposite of this view, it is said, is biological determinism, which implies that humans are constrained by their very minds in what they can think and how they can act.

But, as evolutionary psychologists insist, and as I shall be at pains to show (especially in chapter eight), this all-purpose view of the mind gets the equation exactly backward. The truth is that the more mental modules a mind comes equipped with, the more complex can be that mind's interaction with its environment, and hence the more free can be the possessor of that mind. That, at any rate, is what evolutionary psychologists argue, and what I hope to show. If the evolutionary psychologists are right, then in the future the study of psychology will focus on the structure and operation of the various mental modules that are hardwired into our minds. Of course, evolutionary psychologists stress that our mental modules must be adapted for the specific purpose of providing some survival or reproductive advantage *to our hunter-gatherer ancestors*. But even with this caveat, the representation of the mind provided by evolutionary psychologists is richly complex and diverse.

Indeed, evolutionary psychology is clearly a new psychological and scientific paradigm—in precisely the sense of the term *paradigm* that Thomas Kuhn discusses in his groundbreaking work *The Structure of Scientific Revolutions*.[17] Only time will tell whether this new paradigm represents progress over its competitors. As is the case in all evolutionary processes, if it does represent scientific progress—that is, if it enables greater prediction and control of the natural world than its competitors—it will survive. If it does not, it will simply become extinct. The early work in the field seems encouraging. Like any new and healthy paradigm, evolutionary psychology holds out the promise of providing a feast of new knowledge, if scientists will only get down to the business of working within the paradigm. Each mental module is a main course of study all its own. In that sense, the feast promised by evolutionary psychology is perhaps arrayed on a long buffet table. In the next four chapters I will discuss many of these mental modules at length. But as a way of whetting the reader's appetite, I want to provide some "intellectual hor-d'oeuvres" in the form of a brief discussion of three of what I think are the most interesting mental modules thus far theorized and researched by evolutionary psychologists. Each mental module can be seen as an evolutionary adaptation to a specific selection pressure that would likely have been faced by humans in the ancestral environment. Further, each module can be seen as conferring upon the mind in which it resided a relevant expertise in one arena of interaction with the world, thereby conferring on the human who possessed the mind in question a survival or reproductive advantage that he or she would not otherwise have possessed. With all that said, evolutionary psychologists posit that every (normal) human mind comes equipped with mental modules designed to enable humans naturally to carry out the following information-processing and computational tasks.

Face Recognition. During the 1980s the story spread that when he was governor of California Ronald Reagan had once remarked, "If you've seen one redwood, you've seen them all." Environmentalists were naturally offended. But think how you would feel if you were a redwood. While it may not be particularly important for humans to distinguish two redwoods by sight, it could well be critically important for humans to be able to distinguish other humans by sight. Such an ability is probably no more or less important today than it was in the ancestral environment. It seems we have a mental module specifically designed to do this. Remarkably, the module seems designed not so much to enable us to distinguish faces from other objects, as to distinguish among different human faces. Specifically, we can readily distinguish a particular face, attach an identity to it, store both the general contours of the face and its identity together in our memory, and then recall one attribute when we recall the other, even if the face has been altered by makeup, glasses, hair coloring, and so forth. Evidence for the innateness of this particular module comes from patients who suffer from prosopagnosia. Such individuals can distinguish human faces from balloons or hats, but they cannot distinguish one face—even that of a close relative or their wife or their own face in a mirror—from another random face. Usually patients suffering from prosopagnosia have undergone some type of trauma to the head. One patient, who suffered head injuries as the result of a car accident, lost the ability to recognize the faces of his wife and children, but retained the ability to describe the shape and contour of any given face, and even to determine the sex and relative age of the person to whom the face belonged. But evidence for the innateness of this module was probably clinched by the discovery of an unfortunate individual who, as the result of a head trauma, lost the ability to recognize ordinary objects *other* than human faces.[18]

Mind Attribution. "What was he thinking?" is the most common of questions that we ask when we observe someone has done something we find to be odd or out of the ordinary. We know that other human beings have thoughts and ideas, and that they have emotions like pride and shame, and that these ideas and emotions can change depending upon relatively predictable aspects of the environment. Every normal person knows this because he or she has the subjective awareness of his or her own mental states, and then infers that others must have similar mental states. Babies as young as one year old look in the direction toward which a parent is staring and, even more interestingly, they look directly at a parent's eyes when the parent appears to be doing something novel or odd. Even some primates show a rudimentary ability to attribute minds to other animals. But in humans this ability is manifest and very sophisticated. It could not be otherwise for a species as social as ours. The ability to perceive mental states in oneself, and to infer the existence of such states in others, is today, as it would have been in the ancestral environment, absolutely critical if one is to function well in society.

We can see this by examining individuals who lack this ability. We call such individuals autistic. There are varying degrees of autism, but those who suffer from it to any degree would seem to have at least some malfunction in their mental module for mind attribution. They are, in the words of one theorist, at least partially "mind-blind." There is some evidence to suggest that autistic individuals have difficulty attributing mental states to others because they cannot readily perceive *cognitive* mental states—as opposed to simple physical desires—in themselves. One study found that autistic children made substantially more "pronoun reversal errors"—referring to themselves as "he" or their mothers as "I"—than Down's syndrome or normal children of the same age and level of linguistic development. Autistic children also seem to have substantially more difficulty than normal children in passing what is called the "false-belief" test. In the test, a child is shown a box of animal crackers or some other box or container with which he can be expected to be familiar. But when the box is opened, the child sees that it is filled with marbles rather than animal crackers. Then a second child is brought into the room and shown the same closed box of supposed animal crackers. The first child is asked what the second child will say is in the box. Most children younger than four who happen to be in the position of the first child have some significant difficulty with this test, tending to respond that the second child will say the box contains marbles. The explanation seems to be that young children in the first position cannot attribute a belief that they know to be false to another individual. After the age of four most normal children pass this test without any difficulty. But autistic children still have trouble with this. There is an amazing experiment that seems to clinch the argument that what we are seeing here is a defect in a mental module for *mind* attribution in particular. In this second test, a stuffed animal is taken from inside a bathtub and placed on a sink. A Polaroid picture is taken of the stuffed animal on the sink, and the animal is then placed back in the bathtub. Apparently, children under four think that when the picture develops it will show the animal in the bathtub. Autistic children under four know it will show the animal on the sink.[19]

Spatial "Reasoning." Suppose you are an ancestral human. You have been out hunting all day, and you have managed to roam very far from your home territory. How do you find your way back? In the distance below your level of elevation and slightly to your right you see a familiar landmark that you remember was over your left shoulder when you ventured forth in the morning. You could head directly for the landmark, but that would not quite work. Instead, you form a mental picture of the landscape around you (complete with the landmark), place yourself at your present location in the picture, and then rotate the picture in your mind so that you can figure out precisely in which direction you need to head. Not much has changed in this respect since ancestral days. This is, after all, precisely what you do when you are vacationing in an unfamiliar city, and you are lost, and you happen to have a map showing the location of the landmarks you see around you. To find your way, you simply select a landmark close to you, find the landmark on the map, and then rotate the map (manually or in your mind) until the orientation of the landscape on the map matches the orientation of the landscape around you. Evidence supports the claim that there is a very specific mental module designed to enable individuals to perform mental rotations of objects, and thus to carry out one type of what I am calling spatial "reasoning." Also, there is overwhelming evidence that this ability is extremely highly correlated with skill in higher mathematics. Finally, there is very strong evidence to suggest that after puberty males are significantly better at this activity than are females at any time in their lives. The explanation provided by evolutionary psychologists is simple. In the ancestral environment we can reasonably assume that men did the overwhelming majority of the hunting, and thus men would face a selection pressure for this skill that would not be faced by women.

This may sound like yet another example of sex-based stereotyping. But once again, there is an experiment which seems, if not to clinch this argument, then at least to strengthen it considerably. The experiment did not test the ability of a subject mentally to rotate an object in his or her mind. Rather it tested *another form* of what I am calling spatial "reasoning": the ability of a subject to remember that a particular object was in a given field of objects and to remember where in the field the object was exactly located. In one version of the test, subjects were shown a piece of paper with small drawings of about fifty different objects (a chair, a hat, an elephant, etc.) scattered in an array. After studying that piece of paper, subjects were shown another piece of paper, also with an array of objects. Some objects on the second paper appeared on the first paper, and some did not. Subjects were asked to identify the objects on the second paper that also appeared on the first. Then subjects were handed a third piece of paper, which contained all of the objects on the first piece of paper, but which placed some of these at a location on the paper different from the location at which they appeared on the first piece of paper. Subjects were asked to determine which objects on the third paper had been moved. Finally, in a "naturalistic" version of essentially the same experiment, subjects were lead into a small graduate student office which contained a table, a chair, a lamp, a typewriter, and various other objects placed around the room. Subjects were not permitted to take anything into the room. On a ruse, the subjects were asked to wait in the room several minutes while their "experiment" was being prepared in a laboratory down the hall. After that time, subjects were taken from the room, told the purpose of the study, and asked to name as many objects in the room as they could, and also to indicate where those objects were placed in the room. It turned out that on *all* the versions of this particular test, females outperformed males. In the "naturalistic" setting the female performance advantage was dramatic. Further, when these studies were repeated on females and males of varying ages, the female performance advantage increased by a statistically significant degree after a given female reached puberty. The researchers conclude that the significant female performance advantage on this type of test could be accounted for by the fact that, in the ancestral environment, females faced a selection pressure requiring skill at remembering where objects were located in a confined space, such as a home or homelike environment.[20]

I hope that the extended analysis in the above three examples has served to clarify further exactly what evolutionary psychologists mean when they speak of the various mental modules that come as standard equipment on all (normal) human minds. In this chapter we have explored the answer given by evolutionary psychologists to the question What is the mind? In the next chapter I consider precisely how evolutionary psychologists propose that we study the mind in order to discover and understand its various mental modules. But I cannot leave this chapter without a brief philosophical reflection on the question presented in the title of this chapter, and on the irony of the evolutionary psychologists' response to it.

This reflection begins about a half century ago, when the question on many scientists' minds was not What is the mind? but rather Can computers ever have minds? This second question might, without loss of meaning, be put as, Can machines think? That question was the subject of what remains one of the most provocative papers ever written in the field of artificial intelligence. The paper, entitled "Computing Machinery and Intelligence," was published in 1950 in the journal *Mind*. It was written by Alan Turing, a British mathematician and intellectual polymath.[21] Turing's approach to the question Can machines think? was deceptively simple. He saw the danger in getting caught up in abstract and endlessly contentious arguments about the nature or even the definition of *thinking*, so he proposed an end-run around that whole debate. His proposal was couched in the form of a game he called "the imitation game." Turing spends considerable time in his article discussing the details of this game. I have modified and updated the particulars of the game somewhat. But I have been careful to remain absolutely faithful to Turing's overall point.

Essentially, Turing invites us to imagine a game in which a human interrogator puts whatever type of questions he chooses to two "players" (neither of which he can see), one of which is a human being and the other a computer. The interrogator need not put the same questions to both players. All that is required is that the interrogator have no way of determining whether he is addressing a particular question to the human or to the computer, and that the interrogator have no way of determining—*except by the text of the answer given*—whether a particular answer has come from the human or the computer. Now suppose that we program the computer so that it attempts to give the responses that a human would give to any question it is asked. The interrogator's job is to determine which player is the human. The interrogator "wins" when he makes the correct determination. The computer "wins" when the interrogator makes the wrong determination. What would happen when the interrogator's chance of winning is no better than fifty-fifty? In other words, what would happen when, based strictly on an ordinary conversation, we have no more certainty that we are conversing with a human than we have that we are conversing with a machine? This, according to Turing, helps us to best conceptualize the question Can machines think?[22]

This somewhat modified version of the imitation game that I have just sketched has become known over the last fifty years in artificial intelligence circles as the "Turing Test." This test, and particularly what it implies about the uniqueness of human thought, might appeal to the folks in academic departments like English and communication studies especially, for it suggests that what is distinctive about human thought is not the ability to solve mathematical equations, or even to reason in some abstract way, but rather the ability to carry on a simple conversation with another human. In other words, the ability to use ordinary language, as humans do everyday, is the ability that many of those in the field of artificial intelligence since 1950 have thought forms the clearest line between humans and machines. It is precisely that line that artificial intelligence researchers and engineers have been trying to cross for the last fifty years. Truth be told, they have not yet met with much success at this endeavor. In the various Turing Test "competitions" that have thus far been held, as of this writing (2002), no computer has been able to do very well, even when the subject matter about which the interrogator can converse with the players is drastically limited.[23] So in the strict sense Turing was wrong when he predicted that "in about fifty years' time it will be possible to programme computers . . . to make them play the imitation game so well that an average interrogator will not have more than 70 per cent chance of making the right identification after five minutes of questioning."[24] But we humans should not get overconfident. In the same 1950 article in which he proposes the imitation game, Turing also mentions the game of chess—a game that was once thought to be so complex that no computer could ever play it as well as a human. The game of chess *is* complex. But today, the best chess player in the world is an IBM computer called Deep Blue. Since humans are not evolving very rapidly, but computer programs are, it seems reasonable to say that unless or until genetic engineering brings about smarter humans, no human will ever again beat the best chess-playing computer at that game.

But the ironic point I want to make about the imitation game itself is this: In the 1950s, when some of the cleverest minds in the fields of mathematics and computer science were moved to think about thinking, they thought that for *non*human minds to be able to think, those minds must in some way be able to imitate the behavior of human minds. Today, when some of the cleverest minds in the field of evolutionary psychology think about thinking, they conclude that the best way of understanding how human minds think is to suppose that human minds imitate the behavior of nonhuman minds. Thus human minds become information-processing and computational "machines" with modules that run algorithms. This analogy goes further to suggest that the best way of understanding how the human mind functions is through a process called "reverse engineering." The next chapter examines and critiques that process.

CHAPTER SEVEN

The Challenges of
Reverse Engineering

Rhinoceroses have a big horn at the end of their noses. That may be their most distinctive feature. But they also have thick, coarse, wrinkly hides. How did the rhinoceros get his skin? An author with a Nobel Prize (in literature) provided the following explanation about one hundred years ago: Once upon a time—the author's explanation really does begin with those four words—there lived in the region of the Red Sea a Parsi who loved to bake. One day he baked himself an enormous biscuit. But before he could eat the biscuit an ill-mannered rhinoceros came along, frightened the Parsi, knocked over his stove, and ate his biscuit. The Parsi was not happy. Then, exactly five weeks later, a heat wave struck the area. Everyone took off his clothes to bathe in the sea. The rhinoceros took off his skin, which buttoned underneath, and also went into the water to bathe. In those days, the rhinoceros's skin was smooth and fit snuggly around his body. But as the rhinoceros was bathing the Parsi had an idea. From his house he collected a basket of breadcrumbs—for he never swept out his house. As the rhinoceros was bathing, the Parsi rubbed the breadcrumbs all over the inside of the rhinoceros's skin. When the rhinoceros emerged from the water, he put his skin back on. But, naturally, his skin began to itch. (Here the author explains that the rhinoceros felt as you would if you were lying in a bed that happened to be sprinkled with breadcrumbs.) The rhinoceros scratched, but that only aggravated the situation; he lay down on the ground and rolled around, trying desperately to stop the itching. But all this did nothing but cause his skin to become calloused and wrinkly. Finally, he ran over to a palm tree and rubbed himself against it mightily. But still, this only caused his skin to become that much more calloused and wrinkly. And that is why today rhinoceroses have bad dispositions and calloused, rough, and wrinkly skin.

Rudyard Kipling provided this explanation of how the rhinoceros got his wrinkly skin, as well as other explanations (including one for how the leopard got his spots, and one for how the camel got his hump), in a series of children's stories published at the beginning of the last century.[1] For almost a century, Kipling's *Just So Stories* have been delighting children—and adults who allow themselves to be delighted by such tales. But these stories have also become something of a rhetorical weapon of belittlement in the hands of opponents of evolutionary psychology. Indeed, it is almost impossible to read an account critical of evolutionary psychology without sooner or later coming across the claim that the explanations that evolutionary psychologists offer for whatever human adaptation is currently under study—adaptations, for example, like those for spatial "reasoning" that I mentioned in the previous chapter—are nothing more than "just so stories." This phrase is repeated so often in the anti–evolutionary psychology literature that Kipling's estate should at least get some copyright royalties for its incessant use. I suspect the reason that the phrase is used so often has much to do with its rhetorical power. It provides a nice counterweight to the portentous sounding activity of "reverse engineering" that evolutionary psychologists insist they practice.

Reverse engineering, as the phrase is used by evolutionary psychologists, builds on the metaphor of the mind as a computer—or rather, as an information-processing and computational device. Although the phrase may sound formidable, reverse engineering is simply what one does when one knows the function of a machine but does not quite know exactly how the machine is designed to carry out that function. As Pinker explains, "Reverse engineering is what the boffins at Sony do when a new product is announced by Panasonic, or vice versa. They buy one, bring it back to the lab, take a screwdriver to it, and try to figure out what all the parts are for and how they combine to make the device work."[2] The point to emphasize here is that for reverse engineering to work, one must assume that each and every aspect of the machine under study has a purpose. One must, in other words, take a teleological approach to the artifact in question. Daniel Dennett makes this point clearly in a description of reverse engineering that sounds not unlike the description offered by Pinker. Only the company names are changed:

When Raytheon wants to make an electronic widget to compete with General Electric's widget, they buy several of GE's widgets and proceed to analyze them: that's reverse engineering. They run them, bench-mark them, X-ray them, take them apart, and subject every part of them to interpretive analysis: Why did GE make these wires so heavy? What are these extra ROM registers for? Is this a double layer of insulation, and, if so, why did they bother with it? Notice that the reigning assumption is that all these "why" questions have answers. Everything has a *raison d'être*; GE did nothing in vain.[3]

Note Dennett's last sentence, and recall the quotation in chapter five by Kant, emphasizing that when biologists dissect plants and animals in order to investigate their structure they *must* follow the maxim that nothing in such a creature is *gratuitous*. Of course, because reverse engineers must always embrace this maxim, those who would wish to fool reverse engineers have a powerful tool to use. They could simply design a feature of the machine that has no purpose at all—or rather, whose purpose is to fool reverse engineers into thinking that its purpose is other than to fool reverse engineers. Reverse engineers might spend considerable time and effort attempting to understand the purpose of the given feature. As odd as this might sound, Dennett draws our attention to at least one practical application of this idea. Suppose you wished to construct a fake antique table. After first making the table appear the required age, what if you then drilled a hole right through the left edge of the table? The would-be purchaser of this "antique" might well look at that hole and reason that the table in question must really be an antique, used for some unknown purpose, since no one drills a hole in the top of a table for no purpose at all. Even if the would-be purchaser could not figure out what this hole could possibly be for, the maxim that nothing about the table's design is gratuitous might lead him or her to be fooled by your fake table.[4]

Fortunately for evolutionary psychologists (and biologists), nature cannot attempt to fool those who study it. But this does not mean that reverse engineering is without difficulties for evolutionary psychologists. In fact, there are two distinct but related challenges faced by the theory of evolutionary psychology as it attempts to explain the human mind. The first challenge is to say precisely what survival and/or reproductive problems presented themselves to our hunter-gatherer ancestors. This task obviously involves reconstructing the *very* distant past. The temptation in so doing is to tell "just so stories." The second challenge faced by evolutionary psychologists involves deciding exactly what aspects of the mind constitute discrete features that require an explanation in the first place. As we shall see, this is often more difficult than it sounds. But let me begin with a discussion of the first challenge. When evolutionary psychologists speak of the "adapted mind," they are, as I have repeatedly said, referring to the mind of our hunter-gatherer ancestors. These ancestors lived in what geologists call the Pleistocene epoch. Given the glacially slow pace of human evolution, this epoch accounts for *all* of the *significant* evolution of the modern human brain. Importantly, the time period under consideration here is no more recent than one hundred thousand years before the present. As Pinker explains, "According to the standard timetable in paleoanthropology, the human brain evolved to its modern form in a window that began with the appearance of *Homo habilis* two million years ago and ended with the appearance of 'anatomically modern humans,' *Homo sapiens sapiens*, between 200,000 and 100,000 years ago."[5]

It is important to pause and note the scientific and rhetorical significance of this claim. You might think that this claim supports a bad "biologically deterministic" view of humans. It seems to suggest that we are now *essentially* what we were one hundred thousand years ago. Many individuals, and perhaps many intellectuals on the cultural left especially, seem to want to reject this view because it might imply that modern humans are somehow ill-equipped to meet the demands of modern (read, egalitarian and politically correct) culture. In particular, this view might imply that men are *not* naturally equipped to be stay-at-home dads or that women are *not* naturally equipped to face the competitive pressures of the business world. Of course, this implication would seem to follow only if, in the ancestral environment, men were not likely to be stay-at-home dads and women were not likely to be as competitive as men. These assumptions about the ancestral environment could be wrong, and everything does ultimately hinge on a determination of what selection pressures were faced by our ancestors in the Pleistocene. But, just to hedge their bets, many of those on the cultural left might be tempted at least to embrace an alternative view which held that evolution operates at a much more rapid rate than evolutionary psychologists like Pinker claim.

When compared to the above "biologically deterministic" view, this alternative view might seem preferable, insofar as it can be seen to eliminate the split between ancestral desires and modern duties. We know that the roles of women and men in modern societies have been "evolving" rapidly over the course of the last, say, two thousand years. Indeed, these roles have been "evolving" especially rapidly over the course of just the last two hundred years. As these roles have "evolved," the way in which women and men think about, and relate to, one another must also have changed. If evolution has worked rapidly enough to incorporate these changes into the psyche of modern humans, we could believe that those who resist such changes—those who still think that women have no business in the workplace or that stay-at-home dads are the weak victims of feminist propaganda—represent not the majority but rather the few who have *not* "evolved." They are the "Neanderthals" among us. This alternative view may be quite appealing to those who think that the cultural *right* is populated largely by these unevolved Neanderthals.

But notice how easily this alternative view plays directly into the hands of racists. If evolution really does proceed at a very rapid rate, then while individuals in modern societies might now be well fitted for those societies, individuals who evolved in premodern societies might be *naturally* ill-fitted to live in modern societies. One might call this the Pat Buchanan theory of evolution and cultural competence. During the 1992 presidential campaign, then-candidate Pat Buchanan made the following statement on ABC's "This Week with David Brinkley": "I think God made all people good, but if we had to take a million immigrants in, say Zulus, next year, or Englishmen, and put them in Virginia, what group would be easier to assimilate and would cause less problems for the people of Virginia?"[6]

This little bit of occasional election prose functions as a perfect example of what rhetorical theorists call an *enthymeme*—a purposefully incomplete argument whose meaning can be "constructed" in various ways by various audiences with various viewpoints. Thus liberals might note with no small irony that of course it would be difficult for Virginians to assimilate a million Zulus, because many Virginians are, after all, racists. But Buchanan supporters will hear the quotation differently. They may note the context of the quotation—a discussion of immigration policy in America—and they may nod approvingly at Buchanan's insight that, since evolution proceeds fairly rapidly, it is obvious that those in "less evolved" societies like primitive Africa would have a more difficult time assimilating into contemporary American society than those in "more evolved" societies like modern England. We could therefore expect—Buchanan supporters will conclude—that Zulus displaced to a "modern" society would "naturally" face problems that their "modern" counterparts would not face—problems like finding a job or perhaps even understanding how to live from day to day. Notice that even if you believe that the ancestral environment was an Edenic place to which we should all strive to return, and even if you believe that the society of the twentieth-century Zulus better matches that Edenic ancestral environment than the society of twentieth-century Virginia, you have still conceded Buchanan's point about *difference*. But conceding that point is plainly racist.

Having said this, I grant that there may be some confusion about this example, given the way I am using the phrases "more evolved" and "less evolved." One might argue that, if evolution proceeds at the same rapid rate for all individuals on the planet, then Zulus would be no *less* evolved than "modern" Europeans; they would simply be *differently* evolved. More specifically, Zulus would be especially well evolved to live in the environment of twentieth-century (now twenty-first-century) Africa, and Englishmen would be especially well evolved to live in "modern" Western society. The point, cultural relativists might insist, is that there is no reason to claim one society is "more" or "less" evolved or advanced than another. Zulus may not be particularly good at juggling family and career, as modern society seems to demand, but we "modern" individuals cannot do *their* tricks either.

But I would argue that embracing this view is *still* racist insofar as it concedes the point about difference. That concession alone is all Buchanan supporters need to make their argument. But I suspect that such Buchanan supporters—at least the sophisticated ones—would also wish to go further with the argument. They would wish to argue, based on the factors affecting the speed of evolution (as discussed in chapter five), that Zulus might actually be less evolved than modern individuals, because Zulus faced fewer and less varied selection pressures than were faced by modern individuals, particularly after the invention of agriculture. The point, Buchanan supporters would insist, is that life really is slower and *simpler* for the Zulus.

So it does all come down to a question about the speed of human evolution. Interestingly, one can concede the (perhaps obvious) point that modern Western society is more advanced than Zulu society, and still escape embracing racist views, if one also insists (as do most of the younger evolutionary psychologists) that human evolution is very slow. This would allow one to argue that all of our "modern" advances are built on the foundation of a hunter-gatherer psyche. There is a rhetorical power to this argument. For one thing, it provides a ready explanation for our "modern" problems. They are, in a word, maladaptations. But more generally, this argument allows evolutionary psychologists to insist that humans the world over share the same basic, stable, human nature. Such insistence functions as a sort of rhetorical "insulation" against the charge of racism. Many evolutionary psychologists— and, again, particularly the younger ones—are sophisticated enough to appre- ciate the rhetorical fact that the mere accusation that a particular theory is "racist" is usually enough in modern culture to doom the entire theory. This explains, I think, why evolutionary psychologists are so adamant about the slow pace of human evolution. Thus the editors of *The Adapted Mind* write,

> The few thousand years since the scattered appearance of agriculture is only a small stretch in evolutionary terms, less than 1% of the two million years our ancestors spent as Pleistocene hunter-gatherers. For this reason, it is unlikely that new complex designs—ones requiring the coordinated assembly of many novel, functionally integrated features—could evolve in so few generations. Therefore, it is improbable that our species evolved complex adaptations even to agriculture, let alone to postindustrial society. Moreover, the available evidence strongly supports this view of a single, universal panhuman design, stemming from our long-enduring existence as hunter-gatherers. If selection had constructed complex new adaptations rapidly over historical time, then populations that have been agricultural for several thousand years would differ sharply in their evolved architecture from populations that until re- cently practiced hunting and gathering. *They do not.*[7]

Steven Pinker and Paul Bloom provide empirical support for this claim by looking to the similarity among all human languages. Pinker and Bloom note that "all languages are complex computational systems employing the same basic kinds of rules and representations, with no notable correlation with technological progress: The grammars of industrial societies are no more complex than the grammars of hunter-gatherers; Modern English is not an advance over Old English."[8]

But even if we assume that the mind and psyche of modern Homo sapiens did not evolve significantly during the last one hundred thousand years—let alone the last ten thousand—that still does not answer our original question: Just what was life like back in the Pleistocene? Geoffrey Miller provides a good first approximation of an answer to this question in *The Mating Mind: How Sexual Choice Shaped the Evolution of Human Nature*:

> A fairly coherent picture of Pleistocene life has emerged from anthropol- ogy, archeology, paleontology, primatology, and evolutionary psychology. Like other social primates, our hominid ancestors lived in small, mobile

groups. Females and their children distributed themselves in relation to where the wild plant food grew, and clustered in groups for mutual protection against predators. Males distributed themselves in relation to where the females were. Many members of each group would have been blood relatives. Group membership may have varied daily and seasonally, according to opportunities for finding food and exploiting water sources. . . .

During the days, women would have gathered fruits, vegetables, tubers, berries, and nuts to feed themselves and their children. Men would have tried to show off by hunting game, usually unsuccessfully, returning home empty-handed to beg some yams from the more pragmatic womenfolk. Our ancestors probably did not have to work more than twenty or thirty hours a week to gather enough food to live. They did not have weekends or paid vacation time, but they probably had much more leisure time than we do.[9]

To this account one might add a few more relatively noncontroversial observations. It might be observed, for example, that there were obviously no food storage mechanisms to speak of in the ancestral environment. With respect to animal meat, this meant that what was killed was eaten in a relatively short amount of time. Additionally, there was obviously no glass, plastic, or infant formula in the ancestral environment. This meant that from the beginning of human evolution until less than a hundred years ago, every human being who survived infancy was breast-fed by some female. Sarah Blaffer Hrdy persuasively argues that breast milk was the single most important substance in the ancestral environment. It was essential for the survival of infants, providing not just raw calories but also chemical substances critical for the proper functioning of the immune system; it was required on a daily (in fact, hourly) basis, probably for several years; it could be produced and stored nowhere but inside a woman's body; and without it an infant would die in a few days. Hrdy also notes that the activity of breast-feeding itself may have been the very process that was responsible for the development of the capacity for all types of social relationships in higher species. According to Hrdy, "Sex may not be destiny in the sense that it is necessarily a female that cares for offspring, but lactation requires a female to stay near her young. Prolonged association between mother and suckling young provided both the chance *and* necessity for 'social intelligence' to evolve."[10] Finally, in the ancestral environment there were obviously no contraceptive devices of any kind and no blood or DNA tests for paternity.

What can this knowledge of the ancestral environment bring to our task of reverse engineering the modern mind? The answer is that this knowledge helps us to understand the various functions that our various mental modules were designed to perform, as well as the functions of other, simpler physiological and psychological traits. To see this more clearly, let me take as an example a completely noncontroversial physiological trait that can easily be reverse engineered. Consider the human preference for sweets and fatty foods. We know these preferences are human universals. In a sense, there is very little that needs to be reverse engineered here, because there is no great mystery about the purpose of these preferences. As Pinker notes, they are designed to give us a sensation of physical pleasure so that we are drawn to ripe fruits, which are sweet, and fresh meat, which gets much of its flavor from its fat content.[11] While these preferences are not difficult to reverse engineer, they do draw our attention to an important fact. Preferences that were adaptive in the ancestral environment can often be maladaptive in modern society. Before there were methods for refining sugar and refrigerating (or otherwise storing) meat, no tooth needed to worry about the decay it might suffer as the result of its bearer's consumption of megadoses of processed sugar, and no heart artery needed to worry about the blockage it might sustain as the result of its bearer's access to the abundance of fresh meat that could be hunted down in any modern supermarket with an ease that would have brought tears of joy to even the most manly of our hunter ancestors.

But now let us consider a more complex physiological trait: seasickness. Suppose we attempt to reverse engineer this trait. We might note that a tendency toward seasickness is culturally universal and probably not learned. For it to have developed, it must have conferred some survival or reproductive advantage. An inspection of the ancestral environment will readily suggest what this might be. We can infer that ancestral humans have lived around water—it is, after all, necessary for life. We can further assume, with great confidence, that at least some ancestral humans may have been tempted to venture forth on to that water. Of course, it is one thing to swim in a lake or other large body of water. But to venture far into the water one might require mechanical assistance. Perhaps the first "boat" in the Pleistocene was simply a log to which a would-be sailor clung. Perhaps several logs were somehow tied together to fashion a type of raft. While we cannot know what those first "boats" may have been like, we can know with virtual certainty that they were not very seaworthy. Doubtless many ancestral humans drowned as a result of taking these inferior vessels far out into the water. Thus a selection pressure would be created to stay *off* of these vessels. A trait that caused humans to become nauseous and vomit while on such vessels—and thus ultimately to desire to return to dry land—would clearly have been adaptive in this environment. The fact that we have seaworthy vessels today only shows that many traits that were adaptive in the ancestral environment are maladaptive in advanced societies. At any rate, this explanation certainly demonstrates how reverse engineering can be used to account for a complex and curious physiological trait.

It does indeed. The only problem with this explanation is that it is—to use a word that Pinker employs in similar contexts—pure *bafflegab*. In fact, with the exception of the claim that the tendency toward seasickness is a universal and unlearned trait, there may be nothing in what I have just said that is even remotely true or that makes any sense from an evolutionary perspective. In particular, the claim that ancestral humans were sailing boats *of any kind* is obviously ridiculous. But even if that could be believed, and even if these vessels were poorly designed and therefore likely to sink, neither of these two assumptions would explain nausea and vomiting as an adaptation. Feeling sick to one's stomach while on a boat would not necessarily keep one *off* a boat—at least initially. That goal could be better accomplished by a phobia, of which humans have many. Of course, natural selection need not produce the best adaptive solution to a problem. Still, it is reasonable to ask why nausea should be the manifestation of seasickness, as opposed to headache, joint pain, respiratory allergies, or a hundred other discomforts.

Pinker discusses what may be a *plausible* hypothesis. Seasickness is itself a *subset* of a larger trait: motion sickness. The feeling of nausea that many people experience on board a boat is also experienced by many astronauts, and by many who ride roller coasters. But this insight may not seem to help much, for now we must ask what survival or reproductive benefit would have been gained in the ancestral environment from a susceptibility to motion sickness. None readily suggests itself. Suppose we then take a different approach and ask why *nausea* is associated with motion sickness. Here we might notice that vomiting is certainly a very effective way of ridding the stomach of its contents. Now we have a clue. It turns out that many naturally occurring toxins to which ancestral humans would likely have been subjected act on the body in a very specific way: they disrupt the central nervous system. One effect of this can be to cause the signals coming to our brain from our eyes and our inner ear (which is responsible for balance) to become scrambled. Normally these signals harmonize. We feel that we are walking or running, and our eyes help to confirm that fact. But when one is aboard a boat, particularly when one is in an inside cabin on that boat, the inner ear may well register motion while the eye is convinced from what it sees that the world is stationary. Presumably the most likely way this scrambling could occur in the ancestral environment is by way of some natural toxin entering the body. A well adapted solution to this problem would be to vomit immediately so that the toxin could be removed as soon as possible.[12]

This explanation for seasickness may not be exactly true. But at least it is a sound attempt at reverse engineering. Somewhere between the true but trivial explanation I provided earlier regarding the human preferences for sweets and fatty foods and the first false and fanciful explanation I provided for the human tendency toward seasickness lies the broad expanse of reverse-engineering explanations provided by evolutionary psychologists for all sorts of physiological and psychological traits, including tendencies toward sexual jealousy and sibling rivalry, our ability to use language, our ability to create and to appreciate art, and our preference for certain types of habitats. To successfully reverse engineer these and other traits one needs to know as much as possible about the ancestral environment. That is one challenge facing evolutionary psychologists. A second challenge involves knowing what aspects of the mind constitute discrete features that require an evolutionary explanation in the first place.

To see what is involved in this second challenge, let us return to the example concerning the human tendency toward seasickness. As we saw, our attempt to find an evolutionary explanation for this trait was initially misdirected because we were examining what we thought was a discrete feature of our human physiology that required an evolutionary explanation, but which was in fact a by-product of another feature of our physiology (motion sickness), which was itself yet another by-product of the "real" feature we needed to explain: our tendency to vomit when we have ingested certain toxins. Of course, once you do identify the "real" feature in question, reverse engineering becomes relatively easy. Once one sees that ancestral humans would have needed an ability to survive in an environment full of naturally occurring toxins that attack our central nervous system and cause many of its specific functions to go awry, it is easy to see that the existence of these toxins would have created a selection pressure for developing an adaptive solution to the specific toxins themselves. Vomiting when the input from one's eyes and inner ear do not match would be a uniquely adaptive solution, because it is triggered by the very symptoms produced by the toxin.

The key point is that neither seasickness nor motion sickness as such are adaptive solutions to any problem that ancestral humans would have been likely to face. Importantly, the fact that these two features appear to be "complex" behaviors—they do not *just happen* after all, but must be triggered by very specific inputs—does not negate the conclusion that they are nonetheless by-products of another design feature of the organism in question.

Such by-products abound in the world of man-made artifacts. I once owned a coffee table one leg of which was about an inch shorter than the other three. This naturally created a stability problem that needed to be addressed. My solution was to place a hardcover book (the Loeb Classical Library's edition of Aristotle's *Rhetoric*, in fact) under the shortest leg of the table. It turns out that this edition of Aristotle's *Rhetoric* was extremely well adapted for this purpose, being exactly the right thickness for the job. More specifically, this "feature" of the particular "artifact" in question was very well adapted to be a table stabilizer. But no one would think either that the book itself or its particular thickness were designed for the admittedly odd purpose to which it was put. If one did think this, one would be a bad reverse engineer. But this is not because the physical design of books is unimportant, and all that matters is the words on the page. To be sure, it is almost always the case that the words do matter most. But those who practice the art of book production understand that the design of a book is critical to its ultimate purpose of delivering information. Thus the folks in the production departments of major publishing houses must consider design features like the trim-size of a book, the layout of its pages (which will often determine the number of pages in the book), the quality of the paper on which those pages will be printed, and whether the book is hard or soft cover. In a sense, the thickness of a book is a by-product of these design features. Although books are designed to have standard trim-sizes, they are not generally designed to have standard thicknesses. Thus considering the thickness of a book as a design feature, and attempting to reverse engineer it, is probably not the best approach to take toward this type of artifact. But even that statement needs to be modified. The thickness of a book can be an important design feature if a specific type or amount of text or graphics must be placed on the spine of the book. The point is that in order for reverse engineering to produce meaningful information about an artifact, one needs to know what features of the artifact need to be reverse engineered.

There are numerous man-made artifacts that have been designed for one purpose but that have traits or features that can be adapted for other purposes. Some of these traits and features might themselves seem to be very well designed for the alternative purposes to which they are put. Why then should nature be any different? It turns out that critics of evolutionary psychology have a term for a feature or trait of an organism that *appears* to show evidence of design but that is *not* adapted to the purpose for which it is being used, or perhaps even to any purpose at all. Such features or traits are called *spandrels*.

No book on evolutionary psychology today could possibly be complete without a thorough discussion of spandrels. The term has entered the cultural vocabulary of evolutionary psychologists and anyone who writes about the discipline largely as the result of one of the most interesting and now most often cited papers in the natural sciences. The paper, written by Stephen Jay Gould and Richard Lewontin, and entitled "The Spandrels of San Marco and the Panglossian Paradigm: A Critique of the Adaptationist Programme," was first published in 1979 in the *Proceedings of the Royal Society of London*.[13] Since its publication, the paper has taken on a life of its own, generating an enormous amount of critical reaction throughout the biological and natural sciences, and even spawning a book-length collection of essays solely devoted to understanding the paper from various literary, rhetorical, deconstructionist, and feminist perspectives.[14] The paper's thesis is actually quite simple. Gould and Lewontin argue that evolutionary biologists have become overfocused on adaptation as a way of explaining every feature and trait of a given organism. Although they do not use the phrase *reverse engineering*, Gould and Lewontin argue that there are many important and complex features of organisms that cannot be reverse engineered because these features are simply by-products of some necessary architectural design plan for the organism. Indeed, the term *spandrel* is originally drawn from the world of architecture.

To see what architects mean when they use the term *spandrel* you must first imagine two arches side by side. With profuse apologies to Gould and Lewontin for sullying their image, you are invited to imagine McDonald's "golden arches." If you draw a straight line across the top of those arches, you will have defined a roughly triangular area beneath the line and between the two arches. In architecture, this triangular area is called a spandrel. Now imagine a giant room with four walls. Replace each of the walls with an arch, and place a huge dome on top of the four arches. What you are imagining, if your imagination is sufficiently grand, is roughly the architectural layout of the central space in many medieval cathedrals, including the famous Saint Mark's Cathedral in Venice. Notice that in the space below the dome and between the right angle formed by any two arches you have another roughly triangular area in three dimensions. If you fill in this area with stone or "surcharge" you have a smooth triangular area whose base is the rim of the dome, and which tapers off to a point between the arches. Gould and Lewontin refer to this space as a spandrel also. Actually, they are wrong to do so. Technically speaking, when the arches are at right angles one to another, the triangular space defined between them and below the rim of a dome or ceiling is a *pendentive*. (This came to light, by the way, because Gould and Lewontin's original paper spanned so many disciplines that even architects started critiquing it.[15])

Gould and Lewontin's confusion of spandrels for pendentives is, however, a minor flaw. It does not change their argument, which has much to do with *decorations*. All cathedrals have them, in the form of paintings, statues, stained-glass windows, mosaics, and so forth. As it happens, the four pendentives (or, less accurately, spandrels) that one finds under the giant dome in Saint Mark's Cathedral form ideal spaces for mosaics. And, as you would expect, each of these four pendentives has on it a gorgeous mosaic. The "theme" of each mosaic follows the biblical theme in the painting on the dome itself.

Now suppose you took as a "feature" of Saint Mark's Cathedral in Venice any one of the four mosaics on these pendentives. Suppose further that you attempted to reverse engineer this particular feature of the cathedral. You would notice the extremely elaborate artistic depictions on this particular mosaic. In fact, you would notice how these depictions fit so well within the space formed by the pendentive. You might even be tempted to conclude that the pendentives were constructed to fit the depictions. *But that, of course, would get the situation exactly backward.* On reflection you would realize that the pendentives were constructed first, and the mosaics were made to fit in them. In a sense, then, the mosaics are by-products of the design of the cathedral. Surely the purpose of the pendentives was not to provide a smooth space on which elaborate mosaics could be created. Indeed, the pendentives have no "purpose." They are also by-products of the design of the building. No one designs pendentives. They are simply what you get when you place a dome or ceiling atop a square area formed by four arches. As it is in the architectural world, so it is in the natural world. Many of the "features" of organisms that evolutionary biologists and evolutionary psychologists make the focus of their studies may not be able to be reverse engineered, because they were never "designed" by natural selection in the first place. They are simply by-products of the overall design of the organism. Indeed, there may be very many pendentives—or spandrels—in nature.

That, in a nutshell, is the central argument in Gould and Lewontin's paper. But I readily admit that even though I have summarized the argument correctly, I have not done the paper itself justice. Its persuasive power comes largely from the rhetorical skill with which its central argument is deployed. Gould readily admitted this.[16] John Lyne, a scholar of rhetoric and public discourse who has written extensively on the "construction" of scientific arguments, makes a similar point. Commenting on Gould and Lewontin's paper, Lyne notes that "the rhetoric of the essay is powerful not because it rushes to us with some new empirical finding or theoretical breakthrough but because it *invents* powerful arguments from materials already at hand and presents them in a persuasive way."[17] To give the reader a sense of the persuasiveness of the prose in Gould and Lewontin's paper, let me quote a few passages. The paper opens with the following two paragraphs:

> The great central dome of St. Mark's Cathedral in Venice presents in its mosaic design a detailed iconography expressing the mainstays of Christian faith. Three circles of figures radiate out from a central image of Christ: angels, disciples, and virtues. Each circle is divided into quadrants, even though the dome itself is radically symmetrical in structure. Each quadrant meets one of the four spandrels [i.e., pendentives] in the arches below the dome. Spandrels—the tapering triangular spaces formed by the intersection of two rounded arches at right angles—are necessary architectural by-products of mounting a dome on rounded arches. Each spandrel contains a design admirably fitted into its tapering space. An evangelist sits in the upper part flanked by the heavenly cities. Below, a man representing one of the four Biblical rivers (Tigris, Euphrates, Indus and Nile) pours water from a pitcher into the narrowing space below his feet.
>
> The design is so elaborate, harmonious and purposeful that we are tempted to view it as the starting point of any analysis, as the cause in some sense of the surrounding architecture. But this would invert the proper path of analysis. The system begins with an architectural constraint: the necessary four span-

drels and their tapering triangular form. They provide a space in which the mosaicists worked; they set the quadripartite symmetry of the dome above.[18]

Gould and Lewontin then note that "such architectural constraints abound and we find them easy to understand because we do not impose our biological biases upon them."[19] Clearly—they say—anyone who tried to argue that spandrels (or pendentives) exist so that we can have mosaics in them "would be inviting the same ridicule that Voltaire heaped on Dr. Pangloss: 'Things cannot be other than they are . . . Everything is made for the best purpose. Our noses were made to carry spectacles, so we have spectacles. Legs were clearly intended for breeches, and we wear them.' Yet evolutionary biologists, in their tendency to focus exclusively on immediate adaptation to local conditions, do tend to ignore architectural constraints and perform just such an inversion of explanation."[20] After seizing the reader's attention with some very powerful images, Gould and Lewontin then spell out their argument very plainly:

> We wish to question a deeply engrained habit of thinking among students of evolution. We call it the adaptationist programme, or the Panglossian paradigm. It is rooted in a notion popularized by A. R. Wallace and A. Weismann (but not, as we shall see, by Darwin) towards the end of the nineteenth century: the near omnipotence of natural selection in forging organic design and fashioning the best among possible worlds. This programme regards natural selection as so powerful and the constraints upon it so few that direct production of adaptation through its operation becomes the primary cause of nearly all organic form, function, and behavior. Constraints upon the pervasive power of natural selection are recognized of course. . . . But they are usually dismissed as unimportant or else, and more frustratingly, simply acknowledged and then not taken to heart and invoked.[21]

I hope that these passages provide some sense of the rhetorical power of Gould and Lewontin's paper. In any case, the proof is in the utility and popularity, one might say. On that score, Gould and Lewontin's paper has been enormously powerful. Indeed, it turns out that spandrels are now everywhere in nature. Consider human language, and in particular the complex grammatical structures that can be created in any language. Surely these complex structures—the kind that appear in legal documents and that everyone ridicules, until some smart lawyer finds a lengthy clause that, when carefully parsed, means that you have just signed away your house—could not have evolved in the ancestral environment. What possible selection pressure could have operated to design a mental module for language that would enable me to have written the previous sentence—one that, like the present sentence, is (I would argue) complex and grammatically correct? Since the answer seems to be that there is no such conceivable selection pressure, the overall conclusion must be that language itself is a spandrel. Perhaps it is simply a by-product of an otherwise very complex mind. Indeed, when you think about it (with your very complex mind) you begin to see that a great deal of what the mind can do could not possibly have arisen as an adaptation to any selection pressure in the ancestral environment. What selection pressure could possibly have designed an adaptation that would enable us to solve a first-order differential equation? Mathematics too must be a spandrel. By this logic almost everything we do with our minds would be a spandrel.

But clearly something is wrong here. What has happened is that spandrels have ceased to be a way of telling bad or ill-formed adaptationist accounts from good and well-formed adaptationist accounts, and have instead become a way of saying that we need no adaptationist accounts at all. But spandrels need not be the enemy of adaptation. Consider, again, the example of seasickness as a human trait. It is correct to say that seasickness as such is not an adaptation to any selection pressure in the ancestral environment. In that limited sense seasickness is a spandrel. But an adaptationist explanation can be given for why humans have a tendency to vomit when the input from their eyes does not match the input from their inner ears. So seasickness as such is not inexplicable from an adaptationist perspective. Exactly the same could be said about language. Perhaps it is true that there were no selection pressures in the ancestral environment that would have required humans to speak the way a modern contract law textbook reads—although I am extremely reluctant to grant this point. Humans, after all, place an enormous emphasis on social interaction and social relations. It is quite possible that ancestral humans spent considerable time discussing how the food from which hunt should be distributed to whom under what conditions and when. But even if this were not the case, all it would seem to show is that legalese, deconstruction, and analytic philosophy (to select three complex uses of language today) are spandrels. Language itself would not be shown to be a spandrel. As Pinker and Bloom note, even (perhaps especially) in the ancestral environment "it makes a big difference whether a far-off region is reached by taking the trail that is in front of the large tree or the trail that the large tree is in front of."[22] So there is a selection pressure for what is known as recursion in language. But notice:

once you can say "Take the trail in front of the large tree" you can also say "Take the trail in front of the large tree next to the bush" and you can say "Take the trail in front of the large tree next to the bush on this side of the river." The point is that *even if* language seems more "complex" in modern societies than we could imagine it in the ancestral environment, that complexity is quantitative rather than qualitative in nature. Finally, the argument that higher mathematics is a spandrel can be handled in the same way. It can be conceded that calculus and algebraic topology as such may be spandrels. But they are not just by-products of a very complex mind. They are by-products of a mind that can perform mental rotations of objects and that can "visualize" thought in other ways. As I said in the previous chapter, these abilities seem to be what is essential for proficiency in higher mathematics.

I hope that the discussion in the previous paragraph has helped at least to shore up the view that most human traits, and certainly the complex and interesting ones, can be explained by an adaptationist approach, despite what Messrs. Gould and Lewontin may say about spandrels in their famous paper. To sum up this chapter: There are at least two challenges confronting an evolutionary psychologist who is attempting to reverse engineer the human mind. The first involves knowing what selection pressures were faced by ancestral humans. The second involves deciding which features of the mind are candidates for reverse engineering. When both of these challenges are met with a reasonable degree of success, evolutionary psychologists can tell not "just so stories," but plausible and testable stories about our various mental modules and why they function as they do. There is nothing to say that these stories cannot also be audacious.

I want to conclude this chapter with one such story. I have already mentioned on several occasions Geoffrey Miller's book *The Mating Mind: How Sexual Choice Shaped the Evolution of Human Nature.* As I said, Miller is one of the younger generation of evolutionary psychologists. He is very much into the adaptationist program. Language is no spandrel to him. It evolved as a result of a direct selection pressure. But the selection pressure involved was sexual selection. In chapter five I reviewed how sexual selection works. The classic example, as I said, is the peacock's tail. It grows bigger and more visually stunning because peahens prefer peacocks with big and visually stunning tails. Miller's central thesis is that all our really interesting human traits, especially language and art, can be seen as adaptations to female selection pressure for males who could, for example, speak well and creatively or who could create interesting art. Why would females prefer these males? The answer lies in the handicap principle, which we also saw in the example of the peacock's tail. According to Miller, even if, indeed especially if, the ability to make clever and witty conversation were utterly useless from a survival standpoint—and clever and witty conversations do seem somewhat less than useful when survival is at stake—this ability could still have been crucial from a reproductive standpoint. Males who had this ability in abundance would need to have expended great amounts of mental energy—literally metabolic energy—to develop and sustain the ability. Since the ability did nothing to enhance their survival, it would function as a handicap. Think of the brain as a big peacock's tail. But a male who could acquit himself well in the environment even with this handicap would be signaling his high degree of genetic fitness to potential mates. If you are vaguely sensing that Miller's theory might be able to explain almost all of human intelligence, you are correct. The theory is broad and audacious. But is it true?

Consider, once again, human language. If Miller's theory about sexual selection is correct, we might expect that females have evolved a preference for males who display high verbal skills, while males have evolved such skills. Is this the case? Miller writes,

> As every parent of a teenage boy knows, the sudden transition from early-adolescent minimalist grunting to late-adolescent verbal fluency seems to coincide with the self-confidence necessary for dating girls. The boy's same-sex friends seem to demand little more than quiet, cryptic, grammatically degenerate mumbling, even when playing complex games or arguing the relative merits of various actresses and models. Girls seem to demand much more volume, expressiveness, complexity, fluency, and creativity.[23]

Granted, this may not exactly be hard scientific evidence. But it is suggestive. It also seems to be in line with the work of Cambridge linguist John Locke, who, according to Miller, has studied "the role of 'verbal plumage' in human sexual mate choice." Indeed, Locke draws our attention to a study "in which a young African-American man from Los Angeles patiently explained the sexual-competitive functions of language to a visiting linguist: 'Yo' rap is your thing . . . like your personality. Like you kin style on some dude by rappin' better 'n he do. Show 'im up. Outdo him conversation-wise. Or you can rap to a young lady, you tryin' to impress her, catch her attention—get wid her sex-wise.' "[24] About this Angelino's insights on courtship and language, Miller comments: "In a few concise phrases, this teenager alluded to both classic processes of sexual selection: male competition for status, and female choice for male displays."[25]

But wait. Isn't there a rather significant problem with Miller's whole theory? Women, after all, tend to do better in tests of verbal skill than men. Miller concedes this, frankly noting that

> When sex differences do show up in human mental abilities, women typically show higher average verbal ability, while men show higher average spatial and mathematical ability. For example, women comprehend more words on average, and this sex difference accounts for almost 5 percent of the individual variation in vocabulary size. But sexual selection normally predicts that males evolve larger ornaments. If language evolved as a sexual ornament, it seems that males should have much higher average verbal abilities. Is this a fatal problem?[26]

Obviously, Miller argues that it is not. His argument hinges on the claim that while women may score higher on tests of verbal ability, these tests only measure verbal *comprehension*. They do not measure verbal *creativity*. But, as Miller repeatedly argues in various contexts, creativity is precisely what the men must display to women during courtship. This accounts, according to Miller, for the male, but not female, motivation to produce verbal displays in public and during courtship. He notes that, in comparison to women,

> Men write more books. Men give more lectures. Men ask more questions after lectures. Men dominate mixed-sex committee discussions. Men post more e-mail to Internet discussion groups. To say this is due to patriarchy is to beg the question of the behavior's origin. If men control society, why don't they just shut up and enjoy their supposed prerogatives? The answer is obvious when you consider sexual competition: men can't be quiet because that would give other men a chance to show off verbally. Men often bully women into silence, but this is usually to make room for their own verbal display. If men were dominating public language just to maintain patriarchy, that would qualify as a puzzling example of evolutionary altruism—a costly, risky individual act that helps all of one's sexual competitors (other males) as much as oneself. The ocean of male language that confronts modern women in bookstores, television, newspapers, classrooms, parliaments, and businesses does not necessarily come from a male conspiracy to deny women their voice. It may come from an evolutionary history of sexual selection in which the male motivation to talk was vital to their reproduction. The fact that men

often do not know what they are talking about only shows that the reach of their displays often exceeds their grasp.[27]

Even if you do not buy Miller's theory entirely, it does seem to show that for evolutionary psychologists reverse engineering need not always be a *dull* exercise in cataloging selection pressures and analyzing mental abilities. That said, the next part of the book takes up a very detailed examination of the relationship between mental modules and human culture.

PART THREE

The Nature of Human Cultures

Culture is not causeless and disembodied. It is generated in rich and intricate ways by information-processing mechanisms situated in human minds.

—Jerome H. Barkow, Leda Cosmides, and John Tooby
The Adapted Mind, 1992

CHAPTER EIGHT

The Benefits of Hardwiring

Of all the claims made by evolutionary psychologists, by far the most contentious involves the assertion that the various mental modules that come as standard equipment on all normal human minds are inborn or innate. In the language of computer engineering this amounts to the claim that mental modules are hardwired into us, that they can neither be reprogrammed nor even altered in any way by cultural inputs. This claim suggests—*at least to the critics of evolutionary psychology*—that cultural and societal influences can ultimately have very little impact on human behavior. But this is precisely the kind of biologically deterministic thinking we examined in part one of this book—thinking that is naturally associated with conservative and reactionary politics, but that has always been discredited on careful examination. Thus the critics of evolutionary psychology charge that the theories presented by Steven Pinker and his colleagues are little more than a sophisticated repeat of past assertions that biology is destiny.

I have almost become convinced that evolutionary psychologists actually want their opponents to level this charge against them, since it allows them to respond in a way that enables them to seize the argumentative "high ground," and to accuse their opponents of not keeping up-to-date. Evolutionary psychologists simply note that the issue is no longer one of nature *versus* nurture, or innateness *versus* learning. Rather, they explain, any fool can see that you need *both* nature and nurture, innateness and learning. Hence the claim that mental modules are hardwired into us simply does not equate with the claim that cultural and social influences are unable to affect human behavior. But— the evolutionary psychologists continue—a mind that came into the world without much complexity would actually never be able to get much out of the world. Thus, paradoxically, the more complex the mind is to start with, the more complex culture and society can become, and hence the *less* biology will matter in accounting for human behavior.

To get an idea of how this general argument about complexity works, we can begin with a relatively noncontroversial example. Consider the game of chess. Obviously—evolutionary psychologists might continue—no one is born with the ability to play chess. Thus an adult who has never seen a chessboard is not likely to drop by the southwest corner of New York City's Washington Square Park and immediately beat all of the "regulars" who hang out there playing chess. But presumably anyone can learn to play chess *just because* all (normal) humans are born with complex mental structures that make this learning possible.

And just what are those mental structures that enable us to learn chess? Obviously they include the standard mental *abilities* of perception, attention, memory, and concentration. But I think it is possible to imagine an entity equipped with these abilities but still unable to comprehend the game of chess. Something more—and more complex—is required. To start with, one would need the ability to fix in one's mind the way that each chess piece can move. I am willing to grant that at a *very basic* level memory alone might be all that is needed here. But the situation immediately becomes more complex. Next, one would need the ability to understand the rules of the game. To do so one would surely need the ability to perform fairly complex inferential reasoning. One would need to understand not just how a piece moved, but the conditions under which it *could* move, and even the conditions under which it would be *forced* to move. Hence, for example, one would need the ability to understand that if one's king is in check one cannot move any piece but the king *unless* moving another piece would mean that the king would not then be in check. Beyond that, an ability to "see" different possible chessboard configurations in one's mind would seem to be very close to a necessity. Thus an ability to recognize spatial relationships would seem critical to the game of chess. And to this list could be added many more complex mental structures.

But, finally, since chess is, after all, a game, perhaps a mental structure that enabled one to "get into the mind" of an opponent would be quite useful. On this point it is interesting to note that when the engineers at IBM designed Deep Blue, in addition to programming into it an ability to evaluate configurations of pieces on the board, they also programmed into it an algorithm for evaluating the time between its move and its opponent's response to the move. Since chess matches are elaborately timed, a computer would need to take account of the number of moves it must make in a given amount of time. But the reason for programming the aforementioned algorithm into Deep Blue went beyond this requirement. Presumably, the longer a human opponent took to respond to a given move the more the human was thinking about the move, and hence the better the move was when retrospectively compared to other moves, all other things being equal. Thus Deep Blue had the ability not only to "think" about what move would give it the best position on the board but also to "predict" what types of moves might be most likely mentally to challenge, and possibly even to stump, its opponent. We should not forget that mental concentration requires energy and is tiring to humans. Although Deep Blue did not have to worry about running out of energy during a match, its human opponent did. Humans doubtless pay attention to this aspect of the game, and might also try to "psyche out" their (human) opponents by employing various strategies not directly related to the configuration of pieces on the board. It has been suggested, for example, that Bobby Fischer would sometimes purposefully make a bizarre and apparently inexplicable move just so that his opponent would need to spend precious minutes and precious amounts of mental energy thinking "What has he got up his sleeve this time?"[1]

The point of this example is not to suggest that humans come outfitted with mental modules for playing chess as part of our innate, standard equipment. Rather, the point is that all (normal) humans can learn to play chess because we all have the *various* mental modules necessary for the game. Presumably, one reason the game is so enjoyable is that it draws on, or utilizes, these various mental modules simultaneously. We might even say that the game of chess was invented because we have the mental modules necessary to play it. The *innateness* of these modules is strongly suggested by the fact that people do not need to learn other skills before they can learn chess. Thus chess is not like calculus, which cannot itself be understood without some understanding of algebra. Another way of putting this point is to say that children can learn chess and mathematics and social skills simultaneously. The key to all of this is that innate mental modules do not replace learning; *they direct it.* The more complex the mental module, the more complex can be the learned behaviors.

But what about the claim that the only skill that needs to be hardwired into the human brain to facilitate even very complex learning is a simple ability to *imitate* others? What about the claim, in other words, that all learning is simply imitation directed by, if anything, the simple stimulus-response mechanism that causes us to seek physical pleasure and avoid physical pain? This claim was extremely popular in the middle of the last century. It seems enticing, in part because it has the allure of plasticity. If humans could imitate an extremely wide range of behaviors (as surely we can) this might mean that we could thrive in almost any conceivable social arrangement in which only the most basic biological needs were met—i.e., where we got enough calories per day and enough hours of sleep per night. At the very least, could not just this skill at imitation (assuming it alone came hardwired into humans) suffice to explain complex human behaviors and the cultures thereby produced?

The answer given by evolutionary psychologists is a categorical *no*. A simple facility at imitation is just not enough to account for all of the complex behaviors humans can learn. This is the essence of the point Pinker was making with his contention that many current theories of the mind will go the way of protoplasm in biology since they simply are not up to the task of explaining the complexity of human action and culture.

A comparison of the game of chess with a radically simpler game may help to make this clearer. It is *possible* to imagine someone being able to win (or consistently draw) at the game of tick-tack-toe simply by consistently imitating the moves of an "expert" player. The person doing the imitating would then not need other skills, such as the ability to draw inferences or the ability to manipulate objects spatially in the mind. The person would not even need to understand the game *as such*. Indeed, in 2001, as part of a publicity stunt, gamblers at an Atlantic City casino were invited to try their hands at playing tick-tack-toe against one of several specially trained white Leghorn hens, who were conditioned to "imitate" the moves of an "expert" player. The chickens, who rarely lost—a draw counted as a win for the hen—would peck their own moves on a special board in front of their human opponents and admiring spectators.[2] But it is impossible to imagine that one could learn to play chess simply by attempting to imitate another player who was very good at chess. The combinational possibilities of chess are so great that after only a few moves there would almost certainly be nothing to imitate, since the pattern on the chessboard would very probably be unique to the history of the game. Indeed, it has been estimated that there are 10^{120} legal forty-move chess games— forty moves being roughly the length of an average game.[3] Not even an infinitesimal fraction of these games has yet been played. Thus one could never learn to play chess simply through imitation. For complex activities, learning requires some *preexisting* structure that can guide the learning. Mental modules are those structures.

Ironically, perhaps the best evidence that mental modules *themselves* are innate rather than learned comes from an examination of the skill that almost everyone thinks is the first thing babies learn: language. At first glance, it does not seem as though babies come into the world with anything more than the most rudimentary ability to imitate others. Hence babies must learn language by imitating the language—strictly speaking, the words and combinations of words—used by those around them. For a long time this idea about how babies learned language seemed so obvious that no one bothered to reflect on how impossible the idea is to believe. Then, in the middle of the last century, Noam Chomsky stopped to reflect. What he saw was that the possible number of grammatically correct, say, ten-word sentences is astronomical. And this holds true for any human language. This would rule out imitation as the way to learn language.

To see why, assume that by the time a child is five she has heard on a daily basis an average of only one hundred new grammatically correct sentences. This is a *ridiculously* low estimate, possibly by several orders of magnitude. Remember that a child can hear a new sentence without its being spoken directly to him or her. Children are like sponges when it comes to human language, soaking up sentences from the conversations of adults in the next room, from the mouths of puppets on television, and from the radios in the cars that transport them hither and yon. A half-hour drive to the day-care center in a car whose radio was tuned to an all-news station would doubtless provide a child with at least two or three times the day's quota of new sentences. So my estimate is wildly conservative.

Nonetheless, based on this estimate, by the time the child was five she would have heard close to two hundred thousand different grammatically correct sentences. But what could she possibly do with all of these sentences? Is it even conceivable that she could remember each of these different sentences? The answer is almost certainly no. Remembering two hundred thousand concepts or bits of information in such a way that they can be retrieved quickly is apparently far beyond the capacity of even very highly intelligent individuals. Interestingly, research in cognitive science now suggests that fifty thousand is something like a "hardwired number," forming the upper limit of what people can place in their long-term, but readily accessible, memory about any given area of knowledge. Thus most adults have a working vocabulary of around fifty thousand words or idioms. Interestingly, fifty thousand also just happens to be roughly the number of different configurations of sets of pieces on a chessboard that a grandmaster can store and retrieve quickly from memory.[4] The point is simply that remembering and accessing two hundred thousand different individual sentences would be an impossible task for almost anyone, let alone a five-year-old child.

But even if a child could manage this task, it is not clear what good it would do her. If learning a language meant simply learning to imitate the sentences produced by others, a child of five who was competent in any language *and* who could remember two hundred thousand different sentences and who could retrieve any one of these sentences instantly from memory would still need to find herself in exactly the right context for any one of these sentences to make sense when uttered. This is what would need to happen if children learned language strictly through imitation. But like chess, the "game" of conversation presents individuals with an almost infinite number of combinational choices of well-formed sentences for almost any context. Thus most sentences that children utter are surely not ones that they have heard others say.

We can see this clearly by examining the *mistakes* that children make when they do begin to use language. One can easily imagine a young child uttering a sentence like "I goed to the park." Especially if the child is not around young children, it is extremely unlikely that he would ever have heard this sentence. Thus he cannot be imitating anything he heard. But what he *did* hear is that his mother cook*ed* dinner, his brother play*ed* baseball, and his father watch*ed* television. Honestly, what is a young child suppos*ed* to think? The term *goed* is clearly an inferentially sound construction based on a knowledge of how the past tense is formed in the English language. To be able to construct this term the child would need to have a concept of *the past*; he would need to have realized when he heard the three previously mentioned sentences that the actions of cooking, playing, and watching were all in the past; he would need to have realized when he heard the three previously mentioned sentences that the "basic form" of *cooking*, *playing*, and *watching* was *cook*, *play*, and *watch*; he would need to have inferred that when communicating about a past action one adds *ed* to the appropriate *basic form of the word* for the action; and he would need to have been able to apply this ending to the word *go*. Just describing all of this is complex enough. Yet a child of five could perform all of these mental operations instantly. More astoundingly, an infant put in any human community will learn the language of that community in the same way and at the same rate as any other child who happens to be in a human community that speaks a different language.

From all of this Chomsky reasoned that underlying all human languages is a *universal grammar* that all humans have at birth. The process of learning any particular language, then, is simply the process of "picking up on" the various surface conventions that distinguish one language from another. This can be done fairly easily because, for the most part, these conventions follow regular patterns. Where they are inconsistent—as in *irregular* verbs—mistakes are made until the individual learns the exception by simple memorization.

Chomsky's universal grammar is *essentially* an innate mental module or series of modules for language acquisition. The existence of such a mental module is by now generally accepted, on the basis of overwhelming evidence. Beyond what I have already mentioned, there are fascinating studies involving the transmutation of pidgin languages. A pidgin language is what you get when various individuals with different native languages are thrown together in one place, as was the case on plantations in the American South and the West Indies until about a century and a half ago, and as was the case in Hawaii at around the turn of the last century, when individuals from places as linguistically diverse as China, Portugal, and Puerto Rico were brought to work on the sugar plantations. The first-generation pidgin languages that developed in these communities were obviously ungrammatical hodgepodges of different parts of each of the various languages. The adults could not *naturally* let go of their own different grammatical conventions easily enough to form a good language that they could all share. But then the children came along. Since they had no preestablished *conventions* to unlearn, but did have an innate mental module for language acquisition, the children developed from the pidgin language a shared, new language that was as grammatically rich and structurally consistent as *any* human language. Essentially, they just naturally developed from the mishmash of words and rules they heard an agreed upon vocabulary and a set of shared conventions. *Creole* is the name given to any new language developed in this way. Obviously, Creole could not come about strictly through imitation. It must be the result of the operation of a mental module for language acquisition that is identical in all of us.[5]

The last piece of evidence I will mention in support of the existence of this mental module comes from some fascinating scientific reports concerning one particular family—the "K family," as they are sometimes referred to in the literature—many of whose members suffered a "specific language impairment." Those afflicted with this impairment have significant difficulty in forming grammatically correct English sentences. (All are native English speakers, and all are in the normal IQ range.) The *specific* difficulty these individuals have involves an impairment in their capacity to consistently use tenses properly and in their capacity to construct the plural forms of words. Thus an individual suffering from this affliction might say something like "He was goes to the park today" or "He walked three mile." Given the way this affliction runs in this particular family, there is very strong evidence that it is the result of a single faulty gene. Although this does not mean that there are genes that code for specific grammatical rules—every one of the several accounts of this case that I have read stresses this point mightily—it does strongly suggest that our mental module for language acquisition is coded in our genes and therefore innate.[6]

But suppose that a mental module for language acquisition does exist, and that it is in some relevant ways analogous to Chomsky's universal grammar. I dare say that this fact would not be likely to excite most individuals. To be sure, proper grammar is certainly important to every language user—which means every human. And some individuals—for example those who teach freshman composition in colleges—may be quite passionate about it. But surely it cannot have any deep political significance. Thus even if Chomsky's universal grammar does exist and is innate, one would think this might be of interest only in the scientific and linguistic community. One would be wrong.

To see why, we can begin by noting the special sense in which the term *grammar* can refer either to a particular set of rules for constructing well-formed sentences in any language (and thus for knowing what counts as a well-formed sentence in that language) or to a particular set of rules for constructing anything within any given system (and thus for knowing what is considered well formed in that system). Hence, Chomsky's universal grammar can be seen as a sort of grammar of human grammars. Think of Chomsky as saying that in any well-formed human language there will be a way of expressing past, present, and future action; a way of expressing the object of an action undertaken by someone or something; a way of expressing possession; a way of expressing the modification of some object or state; and so forth. Also, according to Chomsky, all human languages will contain rules for forming all of the expressions that can be formed by any other language. Granted, the rules and certainly the words for expressing something will differ from language to language. But what is expressed will not differ. This point applies within a particular language as well. Thus the sentence "I worked yesterday" is equivalent in meaning to the sentence "I did work yesterday," even though the words in the two sentences are not identical. Similarly, the sentence "Everyone saw the movie" is equivalent in meaning to the sentence "The movie was seen by everyone." Of course, word order does sometimes make a big difference. Thus it matters greatly whether you say "I only have eyes for you" or "Only I have eyes for you." The critical point, however, is simply that while languages differ in the way they express any particular meaning, *the meaning can be expressed in any language.*

We can see this point more clearly by noting that what Chomsky calls *universal grammar* Pinker calls *mentalese.* According to Pinker, mentalese can be defined as "the language of thought."[7] If you understand the term *mentalese* you understand the concept mentalese. Consider this: Steven Pinker is trying to come up with a word to describe the language of thought. By *his* definition this "language" is not spoken in any human community. It is "spoken" in the mind, the realm of the *mental.* But what to call it? Let's see, in Japan they speak Japan*ese*, in Portugal they speak Portugu*ese*, in China they speak Chin*ese* (well, not really, but you get the point). Hence in the mental realm we all "speak" mental*ese*. So Pinker had a concept in his mind—a "word" in mentalese—and he knew the inference rules in English that would allow him to express it. This is how new words, and also new sentences, get formed, from the concepts that exist in mentalese. But notice: *the concepts come first.* Pinker is extremely clear on this point. Mentalese exists independently of any particular language, just as does Chomsky's universal grammar. This suggests that even if you were to ban a word from a particular spoken human language, the concept for that word would still exist in mentalese. Similarly, even if you were to ban a particular grammatical construction from a language—say you banned the use of the suffix *ed*—the concept expressed by the construction would still exist, and people in that language community would simply invent another construction to express the concept, assuming that such an alternative construction did not already exist. (Instead of saying "I worked yesterday" they would just say "I did work yesterday.") If you are dimly sensing that this might have enormous implications for what has come to be called *political correctness,* you would be right.

I often ask my students what they think political correctness entails. Almost all of them respond that it is about being polite when one uses language. Thus it is impolite and politically incorrect to say of someone in a wheelchair that he is a *cripple* or an *invalid* or *disabled* or even a *handicapped person*. The polite and politically correct way to refer to such a person is to say that he is *physically challenged* or, now, *differently abled*. But, surely, if political correctness were just about being polite when speaking to and about other individuals, no one would make such a fuss about it. In fact, political correctness is not ultimately or even primarily about changing people's language. It is about changing people's minds and therefore changing people's behaviors. It is about changing social reality, and it is grounded on the bedrock assumption that the words come first and the concepts that make the reality then follow. Its intellectual grandparents are people like Edward Sapir and Benjamin Lee Whorf. Consider this representative passage from Whorf: "We dissect nature along lines laid down by our native languages. The categories and types that we isolate from the world of phenomena we do not find there because they stare every observer in the face; on the contrary, the world is presented in a kaleidoscopic flux of impression which has to be organized by our minds— and this means largely by the *linguistic* system in our minds."[8]

Pinker notes that individuals who can remember very little about their college education almost certainly remember this hypothesis, or some example that seems to support it. They remember, if only vaguely, that the Hopi Indians have no word for "time" in their language, hence Hopis are able to "stay in the moment" and lead satisfying lives. They remember, if only vaguely, that some primitive tribes have no word for "possession"; hence the members of these tribes lead peaceful, communal existences. I have found that even when the students I teach have not heard of these examples, or others like them, they nonetheless nod in approval when such examples are mentioned. They nod because, in truth, they are hearing something they already know. The idea that language creates reality—social and *perhaps* even physical reality—is as much a part of college education today as is the idea that all knowledge is culturally relative. Both ideas are related in obvious ways. If we dissect and understand our physical and social world based on the words society or culture puts at our disposal, then changing or removing those words would *fundamentally* change our world and ourselves. This bold idea is at the heart of political correctness.

Today there are literally thousands of candidates for words that we need to change in order to improve society. Some of the proposed or already culturally enacted changes are silly—such as referring to *pets* as *animal companions*. But some are not so silly. One idea for change that I find particularly interesting concerns the term *slave*. It has been suggested that American society would be much improved if no one could ever again use this word. Especially with respect to children, the idea is that we should attempt to reform our educational system so that no one will grow up with the ability to say or think something like the following sentence: "Frederick Douglass was once a slave." Those who argue for this change assert that sentences like this are *wrong*, in several senses of that word. Perhaps they have a point. Consider the difference between that sentence and the following one: "Frederick Douglass was once an enslaved person." The difference here may be more than merely stylistic. It may be crucial. In the first sentence, the object of the verb is the noun *slave*. Thus in this sentence the proper noun *Frederick Douglass* is grammatically "equated" with the noun *slave*. In the second sentence the object of the verb is the noun phrase *an enslaved person*, wherein *enslaved* functions simply as an adjective. Thus in this sentence the proper noun *Frederick Douglass* is grammatically "equated" with the noun *person*.

Whether the phrasing actually makes a difference in the way people think is at least debatable. It seems to me that even if the meanings of the two sentences in the previous paragraph are indeed equivalent—and I think that they are—the simple novelty of the construction *enslaved person* does the work of drawing our attention to, and emphasizing, the fact that Frederick Douglass was *always* a person, even though, for some period of time, some people wrongly thought otherwise. Whether the frequent and continued use of the phrase *enslaved person* might eventually decrease the novelty of the phrase, and thereby diminish its "power," is again a debatable question. But notice the way in which the debate over the phrase *enslaved person* reenacts in a curious way the central question of the slavery debates of the eighteenth and nineteenth centuries: the question of whether a slave was a person. Now notice that to be a person one must first be a living thing. Thus, despite what some science-fiction writers might contend, I would argue that we are unlikely to see the day when a future John Brown gives his life for the cause of android freedom. The point I want to make here is a general one concerning all mental modules: we can debate certain distinctions (e.g., whether slaves are persons) just because those distinctions are *not* hardwired into us. Others distinctions we cannot and do not debate, because they *are* hardwired.

Consider, as another example, the abortion debate. I remember once seeing a bumper sticker that read IF YOU DON'T SUPPORT ABORTION, DON'T HAVE ONE. Presumably, the point of this bumper sticker was that abortion can be, and should be, a private matter. But I immediately thought of another "bumper sticker"—one that might have been affixed to the back of the carriage of a nineteenth-century slaveholder: IF YOU DON'T SUPPORT SLAVERY, DON'T OWN SLAVES. Pro-choice advocates will, I am sure, immediately object to this thinly veiled analogy. Slaves were persons, they will insist. But fetuses are not. That, of course, makes the point. The *only* interesting question in the abortion debate is the question of whether fetuses are persons. I am suggesting that we can *sensibly* debate this question because the distinction between fetuses and persons is *not* hardwired into us. But we cannot sensibly debate whether the entity in question is an animate or inanimate object, because *that* distinction *is* hardwired into us. Thus even the most ardent pro-choice advocate would never frame the abortion debate as one involving women and the inanimate objects inside of them, even though the claim that fetuses were inanimate objects, *if accepted*, would seem to end the whole debate. After all, whatever else one may say of it, an inanimate object is not conscious, it cannot feel pain, it is not alive, and hence it cannot be killed. If I am correct, the reason that pro-choice advocates would never frame the abortion debate in this way is that they know that speaking of the fetus as an inanimate object would not be persuasive. I contend that this claim would not be persuasive because the distinction between animate and inanimate objects is hardwired into all humans, and fetuses fall on one side of this distinction in a way that is simply obvious.

If you doubt this, consider the evidence that humans have an innate mental module for what evolutionary psychologists call *intuitive biology*. This mental module enables (or *forces*) us to conceptualize things as either living or nonliving. Research has shown that babies as young as six months make this distinction. To establish this, researchers take advantage of the fact that infants are born with an enormously high degree of *neophilia*: literally, love of novelty. Infants tend to stare intently at scenes that they find strange or difficult to process, while they tend to get bored and look away from scenes they find easy to process. With this in mind, Pinker reports on the following experiment by Elizabeth Spelke.

> A baby is shown a ball rolling behind a screen and another ball emerging from the other side, over and over again to the point of boredom. If the

screen is removed and the infant sees the expected hidden event, one ball hitting the other and launching it on its way, the baby's interest is only momentarily revived; presumably this is what the baby had been imagining all along. But if the screen is removed and the baby sees the magical event of one object stopping dead in its tracks without reaching the second ball, and the second ball taking off mysteriously on its own, the baby stares for much longer. Crucially, infants expect inanimate balls and animate people to move according to different laws. In another scenario, people, not balls, disappeared and appeared from behind the screen. After the screen was removed, the infants showed little surprise when they saw one person stop short and the other up and move; they were more surprised by a collision.[9]

The evidence goes beyond this. As soon as children acquire language, their categories of animate and inanimate seem to snap firmly in place. The key seems to be that living things possess an unchangeable essence while nonliving things do not. Hence other studies find that children are completely willing to believe stories in which a teapot turns into a toaster, but they are much more reticent to believe stories in which a bird turns into a raccoon, and they are adamant that a raccoon could *never* turn into a teapot.[10]

The point is that our innate mental modules tend to direct the debates we have on various issues. Thus from abortion, to affirmative action or racial preferences, to *some* discussions of sexual equality, we are debating about distinctions that are *not* hardwired into us. On other issues, however, there just tends not to be a debate. For example, although there is an ongoing debate about equality between the sexes in America—should women be allowed in combat roles in the military, for instance—there is no *sensible* debate about whether there are any important biological distinctions to be made between men and women. We just know there are. Thus with respect to debates about sexual equality, the question is never one about whether there are any important biological differences between the sexes—a question which, if answered in the negative, would obviously end the debate. The question is, rather, whether these important biological differences matter *in a given context*.

But debates about racial equality may be different. Recall from previous chapters that many scientists—including Richard Lewontin, Edward O. Wilson, and Steven Pinker—insist that there is no important biological distinction to be drawn between humans on the basis of race. It seems entirely reasonable to conclude that while the need to distinguish between humans based on sex is hardwired into at least some of our mental modules, the need to distinguish between humans based on race is not. Thus race itself may be a debatable issue in societies. This would seem to imply that while a color-blind society may be possible, a sex-blind society is not.

On that note, I would now like to suggest a way of applying our entire discussion of Chomsky's universal grammar and Pinker's mentalese to our understanding of mental modules *in general*. In particular, I want to suggest that it would be fruitful to view each of evolutionary psychology's various mental modules as embodying a "grammar" unique to its area of operation or human behavior. Just as the language acquisition module embodies a universal grammar common to all human languages, other mental modules may be usefully thought of as embodying universal "grammars" common to their areas of human behavior. Thus there may be a universal grammar to mate selection, parental investment, exchange between humans, and a whole host of other human endeavors and behaviors.

To see what such a "grammar" might look like with respect to a particular mental module other than the one for language acquisition, let us return to our discussion of the parental investment module. As we saw in chapter six, parental investment in any given child will probably vary based on a wide range of factors related to the child—including health, sex, and birth order— and based on a wide range of factors related to the parents—including certainty of paternity, social status, and overall resource availability. All of these factors are elements of a grammar of parental investment. They are the aspects of the environment that the parental investment module pays attention to as "inputs." Conversely, the parental investment module would not pay attention to other aspects of the environment like, for example, the way in which the past tense of verbs is formed within the language of the given community. A good grammar of mental modules should be able to say something meaningful about the relationship of the elements related to the specific modules.

It turns out that a grammar of parental investment does just this. For example, there is now fairly conclusive evidence that parental investment varies based on the sex of the child and the socioeconomic standing of the parents. There is a "grammatical" rule governing the way this relationship varies. The rule is that the higher the socioeconomic standing of any given set of parents the more likely they are to invest resources in their male children, while the lower the socioeconomic standing of any given set of parents the more likely they are to invest resources in their female children. (In making inferences about parental investment in less developed and/or past societies, researchers look to measures like infant mortality rates, the age at which children are weaned, the spacing between births—which is often related to the age at which children are weaned—and so forth.) This rule is complicated somewhat by the fact that male children in agricultural societies may be more valuable to their parents as laborers than are female children. But this is a factor to be considered when applying the rule; it does not vitiate the rule. Indeed, as Christopher Badcock notes, this rule applies with remarkable precision across vastly different cultures—from strictly caste-based societies like those of nineteenth-century India, to peasant farmers in the small German town of Schleswig-Holstein from the late seventeenth through the nineteenth centuries, even to the "society" comprised of Portuguese nobility from 1380 to 1580. The latter case is particularly striking since, among the nearly four thousand individuals examined, the application of this rule was so precise that it operated even in the context of very fine status distinctions.[11] The evolutionary logic behind the rule is as follows. Because in all cultures hypergyny—the practice of females marrying "up" into the higher socioeconomic group of their husbands—is the norm, it would make sense for low-status families to invest heavily in female children, who could then marry into families of higher status. Conversely, it would make sense for high-status families to invest heavily in males, who would have little trouble finding wives. (Incidentally, the fact that hypergyny *is* the norm in all cultures is itself easily explainable as the outcome of one aspect of mental modules for mate selection—modules that obviously differ between males and females.)

At this point, I would sum up the discussion of the innateness of mental modules by saying that there is ample evidence supporting this claim—with especially persuasive evidence coming from our analysis of the language acquisition module—and that the innate structures of these various mental modules can be usefully analogized as universal "grammars" for various areas of human behavior. Based on this summation it does not seem unreasonable to say, as evolutionary psychologists do, that the innateness and the complexity of these modules provide us with more "choices" than are available to species with fewer or less complex modules. Again, the analogy with language is instructive. A very complex universal grammar allows for an infinite number of combinational possibilities for meaningful sentences in any human language, but each of these meanings is in a sense generated by a grammar that is universal among all humans. Similarly, a very complex parental investment grammar may allow for a very large number of parental investment options in any given society, but each of these options is understandable insofar as it is based on a grammar of parental investment common to the species. Thus individual cultures may vary greatly, but at bottom we are all human beings capable of understanding one another.

That sounds compelling and politically correct. It is also something that I believe strongly. I would therefore like to end this chapter on that note. But I cannot leave this discussion of innateness without mentioning a controversial issue that is almost completely ignored by *most* evolutionary psychologists, but that must logically go hand in hand with any assertion of innateness. The issue involves the slight differences in the *operation* of mental modules from individual to individual, and hence the slight degree to which the mental abilities enabled by these mental modules might be heritable.

We have already seen that, where genetic abnormalities are involved, differences can exist between normal and genetically abnormal humans with respect to the operation of a mental module. Recall the discussion of the "K family," some of whose members were not able because of a genetic abnormality to consistently follow some grammatical rules. It is not inconceivable that, even absent genetic abnormalities, mental modules may differ from individual to individual in the *efficiency* with which they process inputs, and even perhaps, very slightly, in the *structure* of their very design. Thus it does not seem inconceivable that some individuals may be better than others at acquiring language, or at constructing complex grammatical configurations in a given language, or at making "good" parental investment decisions.

But if mental modules can differ at all among individuals, either in the efficiency of their overall operation or in the nuance of their design, and if mental modules are indeed innate, then this would seem to imply that, to some degree at least, these *differences* in mental modules might be *heritable*. This is an issue that virtually no contemporary evolutionary psychologist discusses at any length. This is, I suppose, in keeping with their general aversion to discussing either group or individual differences among humans. The only exception to this ban on the discussion of differences comes with respect to obvious differences between the sexes—differences that are simply too important to ignore. The justification that Pinker in particular gives for this ban—a justification that is at least implicit in *How the Mind Works*—turns on the unarguable point that life is short and there is much to be done, so we ought to tend to important matters first. Getting clear on what mental modules humans possess and how those modules work is, according to Pinker, vastly more important than examining differences between humans in the workings of these modules. Thus of his 1997 book Pinker writes,

> This book is about how the mind works, not about why some people's minds might work a bit better in certain ways than other people's minds. The evidence suggests that humans everywhere on the planet see, talk, and think about objects and people in the same basic way. The difference between Einstein and a high school dropout is trivial compared to the difference between the high school dropout and the best robot in existence, or between the high school dropout and a chimpanzee. That is the mystery I want to address. Nothing could be farther from my subject matter than a comparison between the means of overlapping bell curves from some crude consumer index like IQ.[12]

I can certainly appreciate the argument in this passage and the sincerity with which it is no doubt offered. But I cannot help thinking that there may be something of a rhetorical motive operating here as well. As we saw in part one of this book, any discussion of group differences, especially with regard to race, is nowadays likely to land one in a metaphorical minefield. Pinker and his colleagues have wisely elected not to wander into that minefield and also not to allow themselves to be drawn into it—unless entry is unavoidable, as it is in the case of differences between the sexes. Still, an evasion is not a denial. To the extent that group differences in the operation of these mental modules (other than differences between the sexes) do exist, they will soon enough be found. And if they are found, evolutionary psychology will have contributed, despite the wishes of Pinker and his colleagues, to setting off some mines of its own. It might be better to confront the issue squarely.

Interestingly, among evolutionary psychologists writing popular-science books today, Geoffrey Miller is perhaps the only one who actually says he believes that there are heritable differences in the efficiency and perhaps the structures of some mental modules, and who makes some of these heritable differences the focus of his writing. Yet even Miller does not directly discuss at any length these differences as they relate to race. He does, however, press the point about heritability. Thus in the introduction to *The Mating Mind* Miller writes,

> Evolutionary psychologists Steven Pinker and John Tooby have argued that our science should focus on human universals that have been optimized by evolution, no longer showing any significant differences between individuals,

or any genetic heritability in those differences. That is a good rule of thumb for identifying *survival* adaptations. But . . . it rules out all sexually selected adaptations that evolved specifically to advertise individual differences in health, intelligence, and fitness during courtship. Sexual selection tends to amplify individual differences in traits so that they can be easily judged during mate choice. It also makes some courtship behaviors so costly and difficult that less capable individuals may not bother to produce them at all. For art to qualify as an evolved human adaptation, not everyone has to produce art, and not everyone has to show the same artistic ability. On the contrary, if artistic ability were uniform and universal, our ancestors could not have used it as a criterion for picking sexual partners. . . . [T]he same reasoning may explain why people show such wide variation in their intelligence, language abilities, and moral behavior.[13]

Miller goes on to argue that such reasoning—i.e., sexual selection—*perfectly* explains heritable differences in many important human abilities. Consider our mental module for language acquisition. Miller is very specific in asserting that individual differences in vocabulary size are almost certainly the result of heritable differences in a mental module for language acquisition and/or in general intelligence. Thus he asserts,

> Obviously, vocabulary size differs enormously between people, so it could be a useful cue in mate choice. The American Scholastic Achievement [*sic*] Test includes plenty of vocabulary questions because vocabulary knowledge varies enough to be a reasonable indicator of intelligence and general learning ability. Evidence shows that vocabulary size is at least 60 percent genetically heritable, and has about an 80 percent correlation with general intelligence . . . .
>
> Since words are learned, it may seem odd that overall vocabulary size should be heritable, but that is what behavior-genetic studies find. Identical twins reared apart (who have the same genes but different family environments) correlate about 75 percent for their vocabulary size. By contrast, the environmental effect of parenting accounts for only a small proportion of the variation in the vocabulary size of children, and just about 0 percent of the variation in adult vocabulary size. If you have a large vocabulary, that is because your parents gave you genes for learning lots of words quickly, not because they happened to teach you lots of words.[14]

Earlier I implied that if significant genetic differences are found between groups of humans—and perhaps between races of humans—evolutionary psychologists may have contributed more than they wished to in explaining the reasons for these differences. Sexual selection might be a very powerful explanation. Miller surely realizes this. Hence, like Pinker, he also seeks a way of inoculating himself against the charge that his theory might in any way give aid and comfort to racists. A large dose of that inoculation comes in the following two paragraphs, in which Miller repeats some of the arguments of those like Pinker and Tooby, whom he earlier criticized:

> It should go without saying, but I'll say it anyway: all of the significant evolution in our species occurred in populations with brown and black skins living in Africa. At the beginning of hominid evolution five million years ago, our ape-like ancestors had dark skin just like chimpanzees and gorillas. When modern Homo sapiens evolved a hundred thousand years ago, we still had dark skins. When brain sizes tripled, they tripled in Africans. When sexual choice shaped human nature, it shaped Africans. When language, music, and art evolved, they evolved in Africans. Lighter skins evolved in some European and Asian populations long after the human mind evolved its present capacities.
>
> The skin color of our ancestors does not have much scientific importance. But it does have a political importance given the persistence of anti-black racism. I think that a powerful antidote to such racism is the realization that the human mind is a product of black African females favoring intelligence, kindness, creativity, and articulate language in black African males, and vice versa. Afrocentrism is an appropriate attitude to take when we are thinking about human evolution.[15]

What I said of Steven Pinker earlier also applies here: I can certainly appreciate the argument in this passage and the sincerity with which it is no doubt offered. But I cannot help thinking that there may be something of a rhetorical motive operating here as well. Paradoxically, such passages as this may only serve to fuel the ongoing controversy surrounding questions concerning the benefits of hardwired mental modules. That said, the next chapter takes up yet another controversial question raised by evolutionary psychology.

CHAPTER NINE

What Cultures Can the Mind Run?

Culture is a notoriously difficult term to define. Part of the problem is that, to many who write about it, culture is everything and everywhere. It defines who we are, but in the process defies definition itself. Thus those who would speak of it must fall back on explicit metaphors. Consider the following description of culture, offered by Clifford Geertz, perhaps the most well known cultural anthropologist writing today: "Believing, with Max Weber, that man is an animal suspended in webs of significance he himself has spun, I take culture to be those webs, and the analysis of it to be therefore not an experimental science in search of law but an interpretive one in search of meaning."[1]

For evolutionary psychologists this description of culture will not do, precisely because it places the study of culture outside the realm of the natural sciences. That, of course, is Geertz's whole point. But evolutionary psychologists who write books with titles like *The Adapted Mind: Evolutionary Psychology and the Generation of Culture* and *Darwin, Sex, and Status: Biological Approaches to Mind and Culture* must have a different understanding of culture than do Geertz and his intellectual colleagues.[2] As a way of clarifying what this understanding must be, I want to posit the following question: Suppose that the mind is an information-processing and computational device that comes equipped with various standard, hardwired mental modules. What, then, must culture be if it can be "generated" by such a mind? The answer, I argue, is that culture must be a collection of more or less discrete bits of information that can be acquired and processed as "inputs" and more or less discrete practices and behaviors that can be produced and transmitted as "outputs."

To a large extent, this definition of culture follows from what has already been established about mental modules. While it may not be exactly accurate to speak of minds "running" cultures, this metaphor does draw our attention to some important points—the most important of which may be that cultures cannot generate mental modules. As we saw in the previous chapter, mental modules are the hardwired aspects of the mind. To be sure, an individual obviously cannot acquire or process a culture to which he or she has not been exposed, just as an individual cannot acquire or process a language to which he or she has not been exposed. But this does not mean that culture is in the driver's seat, or the programmer's chair. All it means is that all human cultures that exist today or that have existed can be generated, acquired, processed, and finally transmitted by our various mental modules. But there are some cultures of which the human mind is surely able to conceive that are nonetheless ruled out as *viable* cultures by our mental modules. In other words, there are some cultures—indeed many conceivable cultures—that cannot be run on the evolved architecture of the human mind.

To get a better sense of the relationship evolutionary psychology posits between mental modules and culture I want to consider the various types of cultures that can be generated and acquired by the mate selection modules with which human males and females come equipped. Like parental investment modules, mate selection modules obviously differ between males and females. But they are also similar in some respects. A detailed examination of the female and male mate selection modules will help us see how mental modules in general shape human culture. I should note that, despite its title, the point of this chapter is not to catalog an exhaustive list of all the cultures that the mind can run. Nor is it to generate even a partial list of such cultures. Rather, and more modestly, the point here is to examine at length one set of mental modules with an eye toward seeing how to conceptualize properly the relationship between mental modules and human cultures.

Let us begin our examination with a discussion of the *male* mate selection module. We can infer—and the evidence seems clearly to show—that males are in general significantly less discriminating regarding those with whom they will mate; that over the course of their lives males want significantly more sex partners than females; and that males value youth, health, and attractiveness in a potential mate much more than they value status or control of resources in a potential mate.[3] On all of these points the evidence is in line with what we see in almost all human cultures. Further, the evidence obtained by researchers who actually go out and ask individuals questions related to mate preference tends to confirm the hypothesis that the preferences stated above are, if anything, held in check by cultures, largely because these preferences may conflict with other demands made by other mental modules.

Consider the apparent male preference for a large number of sex partners. Obviously this preference is most easily satisfied if one is not too discriminating about whom he will choose as a sex partner. Thus Pinker reports on one study in which David Buss, an evolutionary psychologist who has studied mate selection modules extensively, sought to obtain information on the amount of time a person would need to know another person before having sex with him or her. Most women said they would probably have sex with an otherwise desirable man as long as they had known him for a year. Most men said they would probably have sex with an otherwise desirable woman as long as they had known her for a week. But that is not the half of it. Most women said they definitely would *not* have sex with an otherwise desirable man they had known for less than a week. What is the corresponding time interval for a man? In other words, what interval of time is such that a man would say, given that interval of time, that he would definitely *not* have sex with a woman? Buss could not find out because, as Pinker reports, his questionnaire did not list a time interval shorter than sixty minutes.[4]

The standard objection to the particular study I have just mentioned, and to others like it, is that men and women are "conditioned" by society to *say* they want—and perhaps even to want—different things from sex. Men are conditioned to say they want many sex partners because that presumably indicates virility. Women are conditioned to say they want fewer sex partners because that presumably indicates chastity. There are of course responses to these objections. One could point out that what individual men and women are telling these researchers seems to chime very well with what evolutionary psychology would predict about male and female preferences with respect to number and quality of sex partners. Further, one could point out that these preferences seem to be culturally universal. Further still, one could point out that the questionnaires given to these subjects were all coded anonymously.

But having said all of this, one would also need to concede that it would be virtually impossible to eliminate all doubt that the subjects in these studies were reporting preferences that were entirely shaped by culture and that could *conceivably* be shaped in the opposite direction. We could all be in the grip of a massive "false consciousness," and simply not know it. Still, the weight of the evidence seems to support the view that these studies are measuring innate preferences. Also, those who argue against these studies face problems of their own. Feminists in particular who argue that present American society is in the grip of a super-powerful false consciousness that clouds everyone's mind with respect to sex roles have a difficult time explaining just how *they* know this. In other words, they cannot explain how their own minds have remained un-clouded. Additionally, individuals who argue that all preferences are "socially constructed" have a difficult time explaining how that argument could be falsified. Consider that, based on any number of measures—from average income, to level of education, to overall socioeconomic status—women in America today enjoy greater freedom and greater equality with respect to men than at any other time or place in human history. Yet American women are the ones who are reporting that they prefer fewer sex partners than men prefer and that they are more choosy than men with respect to those partners. The obvious conclusion would seem to be that these preferences are innate. But to those who would disagree, all of these studies simply serve as that much more evidence supporting the view that sexism must be very deeply engrained in our culture, since even the advances I have mentioned apparently have not served to eradicate it, as evidenced by the different answers men and women still give to questions about the number and quality of sex partners they prefer.

It seems that both sides in this debate could endlessly travel this argu-mentative circle. But with respect to one aspect of the male mate selection module—the aspect that is attuned to attractiveness—we may be able to dis-tinguish socially constructed desire from innate desire in an interesting way. I mentioned that males value youth, health, and attractiveness in a potential mate much more than they value status or control of resources in a potential mate. Youth and health may seem like obvious qualities to value and ones that seem with equal obviousness to be natural as opposed to socially constructed. (Of course this does *not* mean that health is not the result of a person's social position. Clearly it is. Poverty, for example, may make one unhealthy. All this means is that, generally, what is thought of as healthy does not vary from culture to culture. No culture values having cancer.) Taken together, these two qualities—youth and health—may also seem to encompass all that can be said about natural attractiveness. Whatever remained of attractiveness would then be socially constructed and hence a matter of cultural fashion, like hairstyles or preferences for nose or ear rings.

One of these culturally variable measures of attractiveness might be body weight. In the 1990s we were told that "fat is a feminist issue." Part of what this meant was that men's apparent attraction to thin women was not natural (or unnatural) but entirely socially constructed. One popular theory held that the patriarchal power structure linked attractiveness to body weight (a quantity that could easily vary) in order to keep women obsessed about their body weight (and hence their attractiveness) all as part of a conspiracy to distract them from advancing in the workplace and in society in general. There was even proof that low body weight in females did not correlate with attractiveness in all cultures. Part of this proof was to be found on the canvases of the Flemish painter Peter Paul Rubens, whose depictions of what are sometimes described as "voluptuous" women are said to prove—do they not?—that attractiveness is socially constructed.[5]

Actually, they do not. But the issue is complex. It turns out that body weight is not really the key factor in determining a woman's attractiveness to men and even to other women. The key factor is a measurement referred to as *waist-to-hip ratio*. This ratio is around 0.90 for men and for prepubescent girls and postmenopausal women. But among fertile women the ratio is around 0.70. Remarkably, that ratio seems to be a culturally independent measure of the attractiveness of women to both sexes. Because this is a ratio, it can be identical in women of different overall body weights. In fact, one study found that over the course of the last several decades, even as the overall body weight of magazine centerfolds and beauty pageant winners has been decreasing, their waist-to-hip ratio has remained constant at 0.70. Even more remarkably, prehistoric sculptures from more than ten thousand years ago that depict female figures tend to proportion those figures with a waist-to-hip ratio of 0.70, even though many of those figures would be considered significantly overweight by today's standards.[6] The explanation offered by evolutionary psychologists as to the persistence of this ratio over time and across cultures is simple. A 0.70 ratio correlates extremely well with health and *fertility*. Thus this particular ratio would seem to be an important element in the mate selection module of males.

Now let us consider the mate selection module of *females*. As I said, these modules will obviously differ significantly from those of men. But the way they differ is complex. Women's preferences are not just the opposite of men's preferences, although in some cases this may be a useful first approximation. For example, *if forced to choose*, men seem to prefer younger, healthier women with waist-to-hip ratios close to 0.70 whose socioeconomic status is low and who command few resources over older women with waist-to-hip ratios far from

0.70 whose socioeconomic status is high and who command large amounts of resources. On the other hand, women, *if forced to choose*, seem to prefer the opposite: high-status men who command large amounts of resources but who may not be young, fit, or handsome.[7] We will return to women's preference for high-status men shortly. But for now I want to discuss what it means for a man to be handsome.

As you would expect, handsomeness in men, like attractiveness in women, is a fertility indicator. For men, being fertile is associated with the hormone testosterone, which obviously has a number of effects on the body's overall appearance. One striking effect has to do with facial features. When a boy hits puberty, high doses of testosterone produce large amounts of bone mass in the face. If this new bone mass is distributed symmetrically—symmetry being another indicator of health and fertility—the result will be a handsome young man with chiseled features and a prominent, square jaw. Women *should* find this attractive because it is a fertility indicator. But do they?

Here, again, the answer is complex. Some years ago a good deal of media attention was directed toward a study that seemed to show that both men and women preferred male faces with *slightly* feminine features over those with highly masculine features. Feminine features, as you would expect, are largely the opposite of masculine features. Thus feminine features, which indicate female fertility and hence attractiveness, include small, smoothly curved jaws, small noses, and generally less angular bone structures. By the way, the high cheekbones which often signal attractiveness in a woman do not violate this plan of feminine facial beauty. What we see as cheekbones are, in fact, soft-tissue deposits which, given their placement on the face, serve both to highlight the relatively small size of the jaw and nose, and also, possibly, to draw attention to the eyes. Using computer-generated graphics, researchers manipulated the masculine and feminine features of the same face and showed it to subjects in both Western and non-Western cultures. The results seemed to show that almost everyone preferred female faces with highly feminine features, but almost everyone also seemed to prefer male faces with slightly feminine features. Even more strikingly, almost everyone seemed to associate male faces that had highly masculine features with some fairly negative traits like low levels of empathy and trustworthiness.[8]

The problem here is not how we explain men's preferences. Men may prefer feminine faces in either sex just because they always prefer something that looks feminine to something that looks masculine. The problem is how to explain women's preferences. More specifically, the problem is this: Masculine facial features in men are a fertility indicator. Men who have more masculine faces tend, for various reasons, to be more disease resistant and to be in better overall health than men with less masculine faces.[9] But if this is so then the women who preferred less masculine looking men for their sex partners would, by definition, leave less fit offspring than women who preferred as their sex partners men with more masculine features. Hence, a preference among women for men with masculine looking faces would naturally evolve.

Is there any way to resolve the contradiction between what the research seems to show and what nature would seem to dictate? In fact, there is, and the solution is simple. Women, it turns out, want it all. They want men with both masculine *and* slightly feminine looking features. They just want them at different times—specifically, at different times in their monthly cycles. When researchers went back and retested women's preferences, they found that at around the time of ovulation, women preferred pictures of men with masculine facial features, even though these women still rated these men low in the empathy and trustworthiness departments.[10] This may help to explain why smart women make foolish choices. As Sarah Blaffer Hrdy says, "Ovulation can be hazardous to your judgment."[11] But there may be positive ways that cultural conventions can help women avoid some of these "hazards," as we will see when we discuss more specifically what types of cultures mate selection modules can work to shape.

Before getting to that discussion, however, I want to spend a little more time examining the grammar of the female mate selection module. In addition to directing women to pay attention to men's physical features, the female mate selection module clearly directs women to pay *very* close attention to a potential mate's place in the social hierarchy. Women show enormously strong preferences for men at the top of their social hierarchy. In one sense, this preference for high-status men in general is obvious. Men at the top of the social hierarchy tend to command more resources and could therefore be better providers for their children than men at the low end of the social hierarchy.

But what about the claim that women seek high-status men not because of the workings of an innate mental module but rather because the economy pays women on average some fraction of a dollar (estimates vary greatly here) for every dollar earned by men? This is sometimes referred to in the literature as the "structural powerlessness" hypothesis. The argument is that women want men with high status *only* because the women themselves are prevented from commanding enough resources to provide for themselves and their children (and possibly husband).

There are a number of unstated assumptions in this hypothesis. At least one that seems testable is that women who happened to achieve high status in a given culture would tend to act more like men, in the sense that these women would tend to be less discriminating or less choosy about their sex partners and/or potential husbands. But as Bruce Ellis has pointed out, "The structural powerlessness hypothesis is directly contradicted by available data. Interview studies of both medical students and leaders in the women's movement reveal that *women's sexual tastes become more, rather than less, discriminatory as their wealth, power, and social status increases.*"[12]

This could, once again, be the result of some fairly massive societal conditioning. But at the very least, in light of all of the evidence on this point, the burden of proof seems to have shifted to those who would disagree with these findings. It is they who now must identify viable cultures in which women do not generally prefer high-status men, or where high-status women do not prefer even higher-status men. The assumption that such cultures must exist somewhere on the planet is being countered by evolutionary psychologists with the challenge *Show me the culture*. It is not clear that the critics of evolutionary psychology can do this.

Evolutionary psychologists, on the other hand, continue to research the grammars of various mental modules. In addition to what I have already said about mate selection modules, evolutionary psychologists have found differences between men and women in the content of their sexual fantasies (with men's fantasies being more sexually explicit than women's); in the degree to which each sex seeks out and is aroused by any sort of visual depiction of a naked member of the opposite sex (with men being *vastly* more likely than women to seek out and become aroused by such depictions); and in the "cues" regarding sexual jealousy to which men and women respond (with men being more jealous than women of a partner's sexual intercourse with a member of the opposite sex, and women being more jealous than men of a partner's strong emotional intimacy with a member of the opposite sex). Evolutionary psychologists have also discovered *similarities* between men and women in the grammar of their mate selection modules, with both men and women showing equally high levels of preference for potential mates who are deemed *kind* and *intelligent*. And, finally, evolutionary psychologists have uncovered empirical evidence supporting the hypothesis that women are especially attracted toward men who display an openness and affection toward children (which may not be surprising) *regardless* of the women's desire to have children (which may be surprising).[13]

Now that we have examined the mental modules for mate selection we are in a position to ask, How do these modules help to shape the contours of any *viable* human culture? It is precisely this type of question that has made evolutionary psychology such a controversial new discipline in the sciences. I want to try to answer this question by employing what John Rawls in *A Theory of Justice* calls the method of "reflective equilibrium." Rawls develops this method in the context of a discussion of his own highly philosophical theory of justice. But I think the method can be generalized beyond that discussion. Indeed, I would even be so bold as to say that I think this method can provide the much desired bridge that could unite the humanistic critics of evolutionary psychology with the evolutionary psychologists themselves.

The basic idea behind the method of reflective equilibrium is that, in answering a question like "What is justice?" we should begin by considering, on the one hand, all the very specific and highly detailed theories about justice that we can conceive of, and, on the other hand, our own more vague and ill-defined, yet strongly held, intuitions about what would be just in a variety of contexts. We should then move back and forth between our theories and our intuitions, tweaking the theories and rethinking the intuitions, until we arrive at the best possible "fit" between a sufficiently detailed theory and a set of intuitions that are reasonably self-evident. Rawls argues that in using this method we can arrive at an answer to the above question that is "satisfying" to us, in the sense that it fits both the demands of our rigorous theorizing and our own, deeply felt, intuitions.[14]

This method can be fruitfully applied to questions posed by evolutionary psychology. In fact, this method is very similar to what the editors of *The Adapted Mind* are advocating, although they do not call it "reflective equilibrium." To utilize this method in the domain of evolutionary psychology, we would begin with a detailed description of the mental modules we are said to possess and a detailed description of various human cultures. We would then move back and forth between the descriptions of the modules and the descriptions of the various cultures, seeking ultimately to produce the most detailed and useful descriptions of modules and the most complete and accurate accounts of cultures. Sometimes we would need to radically revise our understanding of a given culture because such an understanding could not be fitted with any conceivable design of the relevant mental module or modules. So, for example, even an evolutionary psychologist who had never read Derek Freeman's devastating critique of Margaret Mead's ridiculously inaccurate description of Samoan sexual practices would still reject as unbelievable an account of a society in which status and sex were completely unrelated, in which sexual jealousy of any type was unknown, and in which both females and males adopted the same carefree, casual attitude toward sexual encounters.[15] Such a culture simply could not be fitted with any feasible design for human mate selection modules, given the evolutionary constraints on such designs. On the other hand, there may be some cultural practices that seem viable, if not nearly universal, but that may be difficult to square with descriptions of mental modules for mate selection as they are currently understood. Thus we may need to rethink the design of the modules themselves. I will provide an example of such a cultural practice at the end of this chapter.

For now, I want to apply this method of reflective equilibrium by examining mental modules for mate selection in the light of various more or less universal cultural patterns and practices relating to sex. Let us begin with an obvious cultural practice: the courtship ritual. Given that females are designed to be the choosier sex, we would expect that almost any conceivable human courtship ritual—whatever its specific cultural manifestation—would involve the transfer from a male to a female of something of value (like food) or something that somehow signified the male's high status and access to resources, in exchange for the female's ultimate receptivity to sexual intercourse. Thus among the foraging !Kung San of the Kalahari Desert, animal meat from hunting is a principal element in courtship rituals. Simply put, the better hunters are the better wooers. As Hrdy explains, "the standard !Kung explanation for why a particularly poor hunter remains a bachelor and has no prospects of ever being anything but celibate" was that " 'women like meat.' "[16]

But the exchange in question need not be immediate; nor need it involve something of practical value, like food. The only requirement is that whatever is exchanged be obtainable by males *to different degrees* (so that some males can get more and others less), and that the high-status males obtain more. If the "thing" in question is not actually of practical value, courtship rituals can become all the more interesting. Extravagant gifts of no practical value offered by males in courtship rituals may be an example of the *handicap principle* in operation. Geoffrey Miller makes this argument very convincingly in *The Mating Mind*.[17]

As we have seen, the handicap principle helps us to understand how a survival disadvantage or handicap may actually function as a reproductive advantage. The handicap principle applies throughout evolutionary psychology, wherever there is the possibility of sexual selection. The principle need not apply only to males, but very often it does. Thus a male who wishes to distinguish himself as an especially fit sexual partner may purposefully handicap himself in some way in order to demonstrate that, even with the handicap, he is the equal or superior of rival males. There are any number of ways for a man to handicap himself in modern society. Surely a man who can afford to give a potential mate an extravagant and very costly, *but useless*, gift would be signaling at least two things to her. One thing he would be signaling is that he was able to survive after expending a great amount of resources. The woman might then reasonably conclude that the man really does have a large store of resources to offer her.

Many retail jewelers understand the handicap principle perfectly and are more than happy to assist courting males in putting the principle into practice. Thus one can go to Manhattan's diamond district and purchase a nice 1.5 caret, total weight, princess-cut diamond, with an F color grade and a VS1 clarity rating, for a good price. Or one can walk less than two blocks to Tiffany's and purchase the same stone for four times the price. But Tiffany's will be happy to put your purchase in a special blue box. You know, as does the person to whom you present the item, that the box cost three times what the diamond cost. But *that* is precisely the point.[18]

The unromantic reader may be wondering why any man would shop at Tiffany's. For that matter, why would any man purchase a diamond for a potential mate? If the handicap principle requires only that the man handicap himself by giving up a significant amount of resources, why not give a woman something extravagant, costly, and *useful*, like a sports utility vehicle, or a wide-screen television, or a Nine West gift certificate for the price of the diamond? Would not this demonstrate to the woman that the man in question possessed large amounts of resources and was also *rational*? It might; and that is probably why men do not do this. Recall that in giving an extravagant, costly, and useless gift, the man is sending two signals. The second signal says very loudly that he is, to put it colloquially, a fool for love. It might be quite attractive and rational for a man to demonstrate to a potential mate that his attraction to her is irrational and uncalculating. It might also be important and rational for a man in the ancestral environment to demonstrate this irrationality to others as well. To help make this point, let me use something of an admittedly gruesome example.

Suppose you are an ancestral male human, and a group of other male humans, led by a particularly despicable fellow, has just killed your wife and children. The motives for the murder are not especially relevant. They could include the simple desire to eliminate non-kin who were taking up resources. Your own reaction to the murders is, however, extremely important from the perspective of evolutionary psychology. Since your wife and children have been killed, your prospects for passing on your genes have been significantly reduced, although not necessarily eliminated, assuming you have other surviving family members like brothers and sisters. Still, a reasonable course of action would be to try to obtain another wife, so that you could have more children. In fact, it would be unreasonable to spend time or energy on avenging the murder of your deceased wife and children, since any time or energy you spent in this endeavor would be time or energy you could not spend on searching for another wife and caring for her and your future children. The most reasonable course of action, then, would seem to be to get on with your life and build a nice home for your new family. Indeed, this is exactly what your very rational friends counsel. They tell you that, after all, living well is the best revenge. You would like to believe them. In fact, on a rational level, you do believe them. But about this matter you simply cannot be rational. Something inside of you takes control. It is a kind of madness. Knowing that it will bring certain death, you search out the leader of the group that murdered your wife and children, and kill him (and as many of his cohorts as possible), even though this results in your own inevitable death.

Viewed in one light, this was clearly an irrational action. It might be supposed that individuals who carried genes that predisposed them to act in this irrational manner (i.e., that predisposed them toward a kind of suicidal vengeance) would be selected out of the general population. Their own genes would not survive, while the genes of those more rational males who suffered the same loss but went out and found another wife and had more children would survive. What then can account for the prevalence of irrational vengeance in human society?

The answer may lie in a combination of factors, including sexual selection, courtship cues, the battle of the sexes, and deterrence theory. Notice that as an ancestral *female*, you are on the lookout for a male who will best care for you, and who, if necessary, will risk his life to save you and your children from harm. Even more than that, you want a male who will avenge your murder or the murder of your children, on the eminently reasonable theory that, as a wife of such a male, you would be a less likely target than the wife of a male who would not avenge her murder or the murder of her children. Of course, every male you encounter will say that he would avenge your murder, should such a tragic event come to pass. But whom should you believe? Here, as in much else, *rationality may be the enemy of love*. The more rational a prospective mate, the less desirable he may appear. But the situation is even more complex. If you take an ancestral male's perspective, the best approach would be to try to convince an ancestral female that you would avenge her murder, while secretly intending not to do so. In other words, if you are an ancestral male your best approach is to be a good liar. Of course an ancestral female will know this and so she will seek to develop strategies to ferret out liars. In turn, an ancestral male would know that an ancestral female would know that his best approach was to lie and so he would seek ways to defeat her mechanism for detecting his lying. A kind of arms race ensues in which the stakes are the propagation of one's genes. The situation could go on like this for some time. But what if you were an ancestral male and you could *show* a potential mate that you had rigged your own physiology in such a way that you would *automatically* and *uncontrollably* be forced to avenge her murder, regardless of the consequences to you? The desirability this might confer upon you could be well worth the (rational) control over your actions that you would need to forfeit. Such ancestral males would be signaling to females, *and to other males*, that aggression against them would result in uncontrollable and potentially lethal retaliation, *even if such retaliation were suicidal*. Presumably, such males and their wives and children would be the least likely targets of aggression.

Of course no human, and certainly no ancestral human, could point to particular genes in his body that would predispose him to act in this way. Hence no ancestral female could know *by way of science* that a particular male's physiology was "rigged" for suicidal vengeance in the case of her murder. Still, a perceptive female might be able to get good evidence that the psychological fix was in by observing behaviors that are difficult to fake. For example, the propensity, particularly among many adolescent males, to take enormous physical risks seemingly for no reason at all might be good evidence that they carry the genes that could predispose them toward irrational behavior in general. Coupled with an otherwise aggressive attitude, this might further suggest a tendency toward irrational vengeance, should the context for such behavior ever arise. To an ancestral female concerned for her own survival, the most attractive ancestral male might be one who seemed (within limits, obviously) to be bold, fearless, aggressive to the point of arrogance, and generally *intemperate* in his behavior. Indeed, even today, much to the horror of modern feminists, such "alpha male" characteristics seem to be fairly highly valued by women. To be sure, the protection offered by evolutionarily successful ancestral males may not have been as rational, *or even as effective*, as police forces, electric street lights, small-caliber handguns, Mace, "rape whistles," and self-defense classes—the accoutrements of security in our modern world. But, again, our mental modules are those of our hunter-gatherer ancestors. Hence the ability of a contemporary male to signal to a potential mate that he would give his life for hers, even in avenging her death, might be enough to tip the scales in this male's favor during the courtship ritual.

Courtship rituals are thus clearly connected with status displays *and* with seemingly irrational pronouncements and displays of love. Although their various "vocabularies" may be culturally specific—involving meat, or diamonds, or other items as status displays, and also involving whatever a particular culture counts as risky and daring behavior, as evidence of "irrationality"—their grammar seems universal.

But some may think that, especially in "advanced" societies, such rituals may be purely "cultural," thus having relatively little "real" (that is, *economic*) significance beyond perhaps helping to keep De Beers in business. I would strongly disagree. But even if this *were* so, there is another obvious way beyond structuring courtship rituals that innate differences in male and female mate selection modules work to shape human cultures—a way that may significantly affect the economic arrangements of advanced societies especially. Once again, the choosiness of women with respect to potential mates, and the fact that women tend to base their choices on considerations of status in a potential mate, are the key elements of the equation. Women's choosiness based on men's status means that men are likely to be much more concerned than women about acquiring high status in society. This cannot help but have enormous implications for all types of societies, but particularly for those in which males and females can compete together in areas where status is at stake—areas like the modern "career world."

Earlier I mentioned that some critics of evolutionary psychology insist that women seek high-status men only because these women are prevented from obtaining high status themselves. The misleading assertion that women earn significantly less than a dollar for every dollar earned by a man is offered as evidence for this argument. I mentioned that this "structural powerlessness" hypothesis makes several assumptions. One is that high-status women would not also seek high-status men—an assumption that we saw is wrong. Another more basic assumption is that women *are* actually prevented from achieving high status and the *material* resources, at least, that go with it. The more this issue is examined, the clearer it becomes that this is simply not the case either. Granted, women do earn less than a dollar for every dollar earned by men. But as any number of commentators, including some feminists, have pointed out, this is largely because women choose different, and less materially rewarding, careers than men choose. Or it is because women choose different, less materially rewarding career paths within a given field. Thus, for example, if one compares full-time male and female workers, one finds that males on average work longer hours and spend less time out of work over the course of their working lives than do females. Additionally, females tend to be overrepresented in jobs that pay less but are *more attractive*, as rated by *The Jobs Rated Almanac*, which considers factors like salary, but also, working conditions, opportunity for advancement, safety, job security, and the amount of physical exertion required for the job. Indeed, of the twenty-five *worst* rated jobs, twenty-four had work forces that were over 95 percent male and the twenty-fifth had a work force that was evenly divided between males and females. The obvious conclusion is that women are paid less than men because women choose to work less at more attractive jobs than men.[19] It seems then that women are not prevented by the structure of modern economies from achieving high status and material resources. If anything, modern economies, which reward mental rather than physical labor, provide women with a better opportunity to compete with men than they have had at any other time in history.

Thus the scandal of contemporary feminism is that many feminists have it exactly backward. It is not that women have fewer choices than men in modern American society. Rather, they have more "life choices." In fact, with respect to work and "romance," women have three broad choices in modern American society. They can choose to pursue high status in a career only and thus forsake a potential mate, or they can choose to pursue high status in a potential mate only and thus forsake a career, or they can choose to pursue *both* high status in a career and high status in a potential mate. (If a woman is successful in this third option and children result, these children can always be put in expensive day-care centers or left with high-paid nannies.[20]) On the other hand, men in modern American society can choose to pursue high status in a career only and forsake a potential mate, or they can choose to pursue high status in a career and high status in a potential mate. But *men* cannot choose to pursue high status in a potential mate only and forsake a career, because unless a man is of high status no high-status women will want him. Unfortunately for men, it appears that their options will remain limited. One reason is that powerful, high-status women (many of them avowed feminists, I suppose) who would be in a position to change the culture by marrying low-status males and thus "leading by example" tend, as we have seen, *not* to want to do so. Thus instead of talking about the "glass ceiling," wealthy, well-educated, high-status young women who are feminists might want to consider the following slogan: STRIKE A BLOW FOR EQUALITY. MARRY A POOR MAN.

Beyond courtship rituals and considerations of male and female career patterns, a third way the *female* mate selection module can shape culture can be seen in the approach women take to the difficult mate selection choice I mentioned earlier. Recall the research I described that shows that women, at around the time of ovulation, prefer pictures of men with masculine facial features over pictures of men with slightly feminine features, even though the women rated the men with masculine features low in empathy and trustworthiness. Men with masculine features tend to have better genes—that is, they tend to be more "fit." This presents women with a dilemma, although the exact nature of the dilemma might not initially seem obvious. Rather, it would seem that there is one logical and evolutionarily sound path that women would want to follow. Among the men you are able to attract, select that one with the best—that is, the fittest—genes. In other words, select the man with the most masculine looking face. Such a man will be the fittest, and hence will likely be closer to the top of the status hierarchy, and thus in control of more resources, than other men you could choose. Accordingly, such a man will be the best husband and father. Where is the dilemma?

Although it should not take evolutionary psychologists to point this out, Hrdy and others do note that an especially genetically fit male, and hence one at the top of the status hierarchy, may not always make a good husband or father. Such a male is likely to have gotten to the top of the status hierarchy by possessing traits like aggressiveness and ambition. He might be good at "provisioning" his wife and children, and perhaps keeping them safe from animal predators (of which there are few in the suburbs these days) and from bands of other males (which is now the job of modern police forces), but he does not seem like the type to share the household chores or take junior to soccer practice on the weekends. He is probably working on the weekends. Even worse, if he is a highly aggressive male and in a job that requires a capacity for such aggression (a not unlikely combination), he may have trouble *not* bringing those aggressive tendencies home with him. In one sense, then, a woman might be better off with a genetically fit, high-status male, because of what he can provide in terms of genes and resources. But in another sense, a woman might be better off with a lower-status male (and, thus, we assume, one that is genetically less fit), because of the greater personal involvement he will likely bring to raising children and possibly because of the decreased likelihood that he would display aggressive or violent behavior toward his wife or children. That is the dilemma.

There are a number of approaches a woman could take in the face of this dilemma. She might elect to serendipitously mate with an especially fit male, while nevertheless living with a less fit male, whom she could try to convince that her children were also his. In doing so, however, she and her children would lose the extra degree of resources that the more genetically fit male could provide. (We are assuming here, as before, that the more genetically fit male commands more resources than the less genetically fit male.) Also, even less genetically fit males have mate selection modules that help them to avoid cuckoldry by inducing feelings of jealousy, and parental investment modules which induce them to care for non-biologically related children less than they care for biologically related ones. Thus it might be difficult for a woman to arrange an elicit liaison, and extremely disadvantageous for her and her children if her deception were discovered, since she would lose even the low-status male's resources.

Her second approach might be to mate with and live with the lower-status male, while attempting to teach her male children to be ambitious and aggressive, so that they can make it to the top of the status hierarchy. There may be good reasons for thinking that this could not be done, because temperament (including traits like ambition and aggressiveness) is highly heritable. Also, in doing this the woman would lose the strictly genetic advantages that a high-fitness male could provide, in terms of disease resistance and overall health. But let us forget all of that for a moment, and just think about what would happen if this strategy were successful, in the sense that the woman were able to teach her male children of a low status father to be high-status males. The woman in question would seem to get the best of both worlds. She would be able to mate and live with a low-status male who, because of his low status, would make a good father in the sense that he would be highly involved in the personal lives of his children—i.e., he would be a good "family man"—and she would not need to sacrifice the opportunity that her male children (at least) would grow up to be ambitious and aggressive (thus putting them at the top of the hierarchy), because she could instill these traits into her children through some parenting technique.

Of course, if this woman could gain an advantage by adopting this strategy, then surely *other* women could too. But notice what this would mean. Soon women would be preferring low-status men to high-status men as husbands. Hence it would make sense for a mother to teach her sons to grow up to be low-status males, since that would now increase their prospects of getting a wife and passing on their genes. This is probably the definition of an evolutionarily unstable situation. It would create a runaway race to the *bottom*, as mothers competed to teach their sons to be as unambitious and unaggressive as possible. But a male without any ambition or aggressiveness would not be able to provide for children at all. If all males were like this, no children would be provided for. Hence there would be an incentive for a male to be just a bit more ambitious and aggressive than other males, because his children would be provided for more than others, and hence would survive and reproduce more than others. But there would then be an incentive for a third man to be a little bit more ambitious and aggressive than the second man, and so on, up the ladder of ambition and aggressiveness.

Thus the environment respecting ambition and aggressiveness in males would soon return to the level of the one in which the initial woman in the discussion above faced her dilemma. In other words, the dilemma could not be circumvented by choosing the second approach. However, a third approach might show more promise. What if the woman were to mate with and live with the high-status male, thus getting the benefit of his genes and resources, but somehow teach *him* to be a good family man? What if, in other words, the woman could successfully *domesticate* her ambitious and aggressive husband? By *domesticate* I do not mean that the woman would seek to eliminate traits of ambition or aggressiveness from her mate. Rather, she would seek only to eliminate these traits within the home environment. She might even seek to encourage these traits when they were appropriate, as they probably would be at the male's job. I grant that "compartmentalizing" the trait of ambition in this way might be particularly difficult. An ambitious male at work might be one who needed to *stay* at work for long hours. But I think it possible to "compartmentalize" aggressiveness more successfully. Indeed, if the male in this example happened to be a member of a culture that needed aggressive males (perhaps to defend the culture's homeland as a soldier), then teaching this male how to "compartmentalize" his aggression, so as to display it on the battlefield but not in the home environment—including the "home" environment of the town streets—would benefit not only the male's spouse but the culture at large. We might expect, then, that the culture at large would attempt to help this male's spouse domesticate him, perhaps by attempting to instill in this male a "code" that taught him to be aggressive on the battlefield but gentle and mild tempered at home and in the culture at large. If the culture at large just happened to be the one that existed in Europe during, say, the twelfth through fifteenth centuries, it might call this code *chivalry*. A code of chivalry—or something like it—would certainly "fit" with a female mate selection module that attuned women to be attracted to males who were ambitious, aggressive, and always ready to defend them, yet also gentle, respectful, and even refined. Perhaps this survives in other cultures as female preferences for "an officer and a gentleman."

I think it fair to say that the various cultural patterns and practices I have sketched, from courtship rituals to workplace status to chivalry, are consistent with what evolutionary psychologists tell us about the structure of both male and female mate selection modules. In other words, there is a good "fit" between the modules and the cultural patterns and practices. Thus if we applied the method of reflective equilibrium described earlier to the cases thus far examined we would have a high degree of confidence that we have understood the workings of the appropriate mental modules correctly and that we have accurately described the relevant cultural patterns and practices.

But I promised to provide an example of a cultural practice that seemed not to fit too well with what we think we know about male and female mate selection modules. The example that I have chosen is one that I came across while reading Richard Dawkins's 1976 book *The Selfish Gene*. The "problem" that Dawkins identifies between our mate selection modules and one of our cultural practices was certainly recognized before 1976. And Dawkins himself does not use the exact language of evolutionary psychology. He does not speak, for example, of mental modules. But he does put the problem very lucidly. Thus I want to quote directly from his book. The passage that follows reproduces the last two paragraphs of a chapter from *The Selfish Gene* entitled "The Battle of the Sexes." In the chapter, Dawkins examines the reproductive conflicts of interest between the sexes among various species, and he attempts to demonstrate how mating strategies have evolved to deal with these conflicts. He concludes the chapter with a very short discussion of the human battle of the sexes, noting, as I have, that women have more "invested" in the sex act than men, and suggesting, as I have also said, that this should mean that women are more choosy about their sex partners than are men. So far, fine. But this choosiness on the part of women should mean that men are the sex that *advertises* its fitness. Men do, of course, advertise such fitness in athletic displays and in other competitions between males. But what about direct advertisement in the way of body ornamentation? What about clothes, hair, and makeup? This is how Dawkins explains the problem:

> One feature of our own society that seems decidedly anomalous is the matter of sexual advertisement. As we have seen, it is strongly to be expected on evolutionary grounds that, where the sexes differ, it should be the males that advertise and the females that are drab. Modern western man is undoubtedly exceptional in this respect. It is of course true that some men dress flamboyantly and some women dress drably but, on average, there can be no doubt that in our society the equivalent of the peacock's tail is exhibited by the female, not by the male. Women paint their faces and glue on false eyelashes. Apart from special cases, like actors, men do not. Women seem to be interested in their own personal appearance and they are encouraged in this by their magazines and journals. Men's magazines are less preoccupied with male sexual attractiveness, and a man who is unusually interested in his own dress and appearance is apt to arouse suspicion, both among men and among women. When a woman is described in conversation, it is quite likely that her sexual attractiveness, or lack of it, will be prominently mentioned. This is true, whether the speaker is a man or a woman. When a man is described, the adjectives used are much more likely to have nothing to do with sex.
>
> Faced with these facts, a biologist would be forced to suspect that he was looking at a society in which females compete for males, rather than vice versa. In the case of birds of paradise, we decided that females are drab

because they do not need to compete for males. Males are bright and ostentatious because females are in demand and can afford to be choosy. The reason female birds of paradise are in demand is that eggs are a more scarce resource than sperms. What has happened in modern western man? Has the male really become the sought-after sex, the one that is in demand, the sex that can afford to be choosy? If so, why?[21]

There are a number of approaches one can take toward answering Dawkins's question. One could note that in almost all areas of culture (from sports to politics) men "advertise" their fitness by competing, specifically against other men. But "almost" is not good enough in evolutionary theory. If men could gain *any* reproductive advantage at all from advertising their fitness by dressing flamboyantly or wearing makeup (which could enhance the appearance of their health, for example), evolution would select for men with a desire to do just this. Again, one needs to answer why it is in women, and not men, that this desire seems to predominate.

One could next respond that men do, in fact, care about their appearances. There are, for instance, a wide range of products directed at men which promise to get rid of gray hair, or to help men grow more hair (on their heads), or to otherwise enhance a particular aspect of a male's appearance. Perhaps these products represent the wave of the future. In her book *The Beauty Myth*, Naomi Wolf argues that the modern "beauty industry" will soon be targeting men with the same zeal that it leveled "against" women.[22] This may be true, and the beauty industry may ultimately be successful in its attempts to sell a very wide range of cosmetics to men. If so, we might conclude that, in the matter of cosmetic usage among the sexes and perhaps also in the matter of dress, mate selection modules allow for an unusually large degree of cultural variance.

Finally, one might attempt to flatly contradict Dawkins's whole argument by insisting that, with respect to men and women, the manner and extent of their advertisement in regard to dress and (perhaps) cosmetics usage is what evolutionary theory would indeed predict. This point may be easiest to make with respect to clothing. There may be some truth after all to the old adage that men get dressed up by putting on clothes while women get dressed up by taking off clothes. Jennifer Lopez's appearance at the 1999 Grammy Awards in a dress that, as the critics said, left almost nothing to the imagination, is but an extreme example of this phenomenon. Perhaps the real utility of the little black dress is that it is, in fact, *little*. For women, less is more. For men, on the other hand, more ornamentation (at least in the form of clothing) is required because women need to know not only that a given person is a male, but also what other attributes besides simple maleness he might possess. Clothing that allowed males to show off their place in the status hierarchy would fit this requirement. Thus it may be no accident that we say of a soldier in a uniform bedecked with many medals that he is "highly decorated."

But I am not sure this argument works either. It is not clear why men's clothing could not be designed to show off much of the male body while also allowing sufficient space to wear medals and other ornaments. It seems, therefore, that the relationship between the male and female mate selection modules and the usage of clothing and cosmetics by males and females (in Western culture at least) may not be in reflective equilibrium quite yet. Perhaps this is a fruitful area for further research.

In this chapter, I have tried to suggest some ways in which some mental modules can, and perhaps cannot, account for various cultural practices. In the final chapter, I examine at length what I think is the most interesting mental module of all.

CHAPTER TEN

The Evolutionary Psychology of "Little House on the Prairie"

From 1974 to 1983, millions of American families turned on their television sets every week to watch the Ingalls family struggle with life on the nineteenth-century American prairie. Indeed, during its nine-season run the show "Little House on the Prairie" became one of the highest rated dramas on television. Each episode was a kind of sixty-minute—well, fifty-two minutes with commercials—morality play wherein we would learn a useful life lesson. Sometimes we would learn that the best gift is one that comes from the heart; sometimes we would learn that running away never solved any problems; and sometimes we would learn that practical experience can be just as valuable as book-smarts. But always the same general message was conveyed on each and every episode. That message was devastatingly simple: that life on the prairie—or anywhere else for that matter—works best when helpful people help other helpful people. In a now famous article, published in 1971, the proto-evolutionary psychologist Robert Trivers came up with a name for what made that best life possible on the prairie. He called it *reciprocal altruism.*[1]

To my mind, the general idea of reciprocal altruism, and all of the various research and theory that has accreted around it, represents some of the most fascinating work in the field of evolutionary psychology to date. As we shall see, the theory of reciprocal altruism is fascinating because it nicely integrates evolutionary psychology with some critical concepts in economics and game theory; because it carries some startling implications for the *content specific* nature of the process of human reasoning; and because it can account extremely well for distinctly political emotions, including our sense of justice and fair play.

To begin our analysis, we need to get clear on exactly what is, and is not, meant by the phrase *reciprocal altruism*. Here we will do well to keep in mind the following lament offered by Trivers: "Models that attempt to explain altruistic behavior in terms of natural selection are models designed to take the altruism out of altruism."[2] What did Trivers mean by this?

Consider the behavior of a parent who works twenty hours of overtime a week so that he can afford to send his (biological) child to an expensive private elementary school, rather than the neighborhood public school. We might call such behavior laudatory. But, according to Trivers, we should not call it altruistic. Clearly, by benefiting his child this parent is benefiting himself, by benefiting that half of his genes that are inside of his child. This is an example of what is known in evolutionary biology as *kin selection*. The idea is simple. Individual genes could replicate very quickly if they built bodies that aided other bodies that contained high percentages of the genes in question. Hence, we would expect humans to come equipped with an especially strong desire to aid their kin. It turns out that we even have a mental module that is apparently dedicated to enabling us to notice and keep track of the type and degree of relatedness of others to ourselves. This is exactly what we would expect from "selfish" genes.

But for just this reason we may *not* want to encourage throughout society the type of parental behavior mentioned above. We may instead want to frankly call such behavior selfish. Do not all major religions tell us that everyone is a child of God? Should we not therefore have as much love for a random child as we have for our biological child? Should we not encourage—or perhaps compel—the parent in this example to take that extra money he has made from working overtime and give it to the neighborhood school so that all children (including his own) will benefit, at least to *some* degree? These are all philosophical questions. But they also have a direct relevance to the concerns of this chapter. For now, however, I want to emphasize a more specific biological point that Trivers makes very clearly at the beginning of his article. The point is that although there are numerous evolutionary models that can account for apparently self-sacrificing behaviors directed toward kin, we should not call such behaviors altruistic precisely because they *do* confer an evolutionary advantage on the person engaging in the behavior. They help reproduce *some* of his genes. More specifically, they improve his *inclusive fitness*—that is, they improve the quantity of his own fitness plus the fitness of his kin discounted by their degree of relatedness to him. Throughout this chapter, I will follow Trivers in not calling such types of behaviors altruistic.

As it turns out, this exclusion rules out quite a lot of what we might consider altruistic behavior. In particular, it rules out cases in which individuals *deliberately* sacrifice themselves for a community without any possibility of obtaining *for themselves* the additional resources that might have followed from the high status conferred by the sacrifice. The behavior of Japanese kamikaze pilots during World War II would be an example of such a sacrifice. So too would the apparently much different behavior of individuals who joined a highly esteemed priesthood that required celibacy from its members. (Evolutionarily speaking, deliberately remaining celibate for a cause is of course the same as sacrificing one's life for the cause.)

But both of these self-sacrificing behaviors (suicide in battle and a priestly life of celibacy) might be explainable as examples of kin selection if the high status they conferred on the self-sacrificing individual tended to be transferable to the individual's immediate family. And, in fact, in all human societies status is easily transferable from oneself to one's kin. So the ultimate sacrifice of an individual's life might result in the greatest number of his genes surviving into future generations if that sacrifice confers more status on his family than the status his otherwise mediocre life would have conferred on him (and his family).

Viewed in this light, any number of seemingly altruistic acts become examples of selfish genes at work. If this is beginning to sound extremely cynical, you may perhaps be appreciating the wisdom of Trivers's comment that many evolutionary models do tend to take the altruism out of altruism. Once again, it is critical to emphasize that for Trivers kin selection is not an example of altruism precisely because, *by itself*, kin selection cannot reach beyond the family. But we *want* to extend the kind of unselfishness we find in families to the larger community. As I said, all major religions tell us that all men are "brothers." It is no coincidence that all societies have a term for the kind of unselfishness that transcends the family. Some call it camaraderie, some call it fraternity, some call it collegiality, some call it solidarity, some call it mutual cooperation, some call it neighborliness. Trivers calls it reciprocal altruism. We want such behavior to be the norm in society. But, unfortunately, mutual cooperation is not always in everyone's rational self-interest. We can see why by examining something known as the Prisoner's Dilemma—probably the most famous "game" in the field of game theory.

Just as no book on evolutionary psychology could possibly be complete without a discussion of spandrels, no such book could be complete without also a discussion of the Prisoner's Dilemma. We owe the discovery—or at least the mathematical formulation—of the Prisoner's Dilemma to a pair of RAND Corporation researchers, Merrill Flood and Melvin Dresher. It was Flood's 20 June 1952 RAND research memorandum entitled "Some Experimental Games" that brought the Prisoner's Dilemma within the ambit of the then rapidly developing field of game theory. Since then the dilemma has been the subject of innumerable books and articles, many of which have tried (with perhaps only mixed success) to figure out exactly what the dilemma actually means for humanity.[3]

The Prisoner's Dilemma obviously need not involve real prisoners. As long as the "payoff" matrix is correct it can involve any two discrete players. For the sake of tradition, the following example does focus on prisoners. I have, however, embellished the presentation of this idea considerably. Suppose that you and a cell mate are stuck in Folsom prison. Both of you are serving two-year sentences. The guards in Folsom prison are capable of administering beating to the inmates, ranging from mild, to medium, to severe. Of course, the guards can also administer no beating. These three levels of beatings, plus the condition of not being beaten, will function as our payoffs.

I need to pause here for a moment for a quick stipulation. Most Prisoner's Dilemma "games" use payoffs that can be readily *quantified*. I have used levels of beatings as payoffs. But even though the level of discomfort suffered in a beating may not be quantifiable, a Prisoner's Dilemma can still be generated using these payoffs. In fact, the example I have chosen is probably the absolute simplest form of the Prisoner's Dilemma. Thus we will agree to the following: For every prisoner in Folsom prison, a severe beating hurts more than a medium beating, which hurts more than a mild beating, which hurts more than no beating at all.

Now to continue. Every morning for the last two years, between the hours of eight and nine o'clock, Bruiser, the guard on your cellblock, has entered every cell on your block and administered a medium-level beating to each inmate. Fortunately, tomorrow is the last day in prison for you and for your cell mate. At noon tomorrow both of you will be free men. Still, both of you are naturally tired of the beatings. Both of you also know how the prison system works. In fact, everyone on your cellblock knows that if *both* inmates *in a given cell* accuse Bruiser of mistreatment (that is, beatings), Bruiser will be reprimanded, removed from the cellblock, and replaced with another guard named Peewee. Everyone on the cellblock also knows that Peewee administers only mild-level beatings every morning. Also, everyone on the cell-block knows that if only one of the two individuals in a given cell accuses Bruiser of mistreatment while the other refuses to corroborate the accusation, then Bruiser will not be reprimanded or removed. Further, in such a situation, Bruiser will retaliate against the accuser by beating him severely the next morning, and Bruiser will reward the nonaccuser by not beating him at all the next morning. Knowing all this, you and your cell mate nonetheless agree that after supper both of you will request to see the warden. At that meeting each of you will then accuse Bruiser of mistreatment.

Curiously, shortly before supper, Bruiser replaces your cell mate with another individual who will also be released tomorrow at noon, and whom you know to be on the cellblock, but whom you have not seen previously. (Bruiser is not stupid. We will discuss this move shortly.) In the five minutes you have to speak to him before supper—during which time there is no talking allowed—your new cell mate tells you that *his* old cell mate (who is also scheduled for release tomorrow) and he agreed just that day that immediately after supper both of them would request to see the warden and then accuse Bruiser of mistreatment. You and your new cell mate agree to do just that after supper.

Immediately after supper, Bruiser comes to your cell. Before he opens the cell door he announces that each of you will be taken separately to see the warden and that after the first prisoner has seen the warden that prisoner will be placed in solitary confinement until the second prisoner has seen the warden. Bruiser then escorts your cell mate to the warden's office. You wait patiently in your cell. Presently, Bruiser returns and escorts you to the warden's office. You and Bruiser enter the warden's office, and Bruiser closes the door behind you. It is just the three of you in the office. The warden looks up from his desk. "You wanted to see me?" he asks. At that moment you have one of two choices. You can accuse Bruiser of mistreatment or you can complain about the food in the cafeteria. Which do you do?

Let us further stipulate that you are a completely rational and entirely self-interested individual. Let us finally stipulate that your interest is to have the best go of it as you can tomorrow morning. You realize that tomorrow morning you *will* receive *either* no beating, a mild beating, a medium beating, or a severe beating. You, and all other rational inmates, prefer the first outcome to the second, the second to the third, and the third to the fourth. Again, what do you do?

Under these conditions, game theory says that you should—indeed, as a rational, self-interested agent, you *must*—reason as follows: My new cell mate had only one of two options. Either he accused Bruiser of mistreatment or he did not accuse Bruiser of mistreatment. If he did *not*, I *must not*; otherwise I will end up being beaten severely, when I could end up only being beaten at a medium level, as usual. On the other hand, if my new cell mate *did* accuse Bruiser of mistreatment, I *must not*; otherwise I will end up being beaten mildly (by Peewee), when I could end up not being beaten at all (as reward from Bruiser). Hence, *regardless* of what my new cell mate did, I *must not* accuse Bruiser. Of course, your new cell mate, being an equally rational and equally self-interested individual, who has the same interest as you do in having the best go of it tomorrow morning and who understands the system as well as you do, will reason as you did, and conclude that he *must not* accuse Bruiser. So both you and your new cell mate complain about the cafeteria food. This means that both of you, acting entirely rationally and on the basis of your own individual self-interest, will receive a medium-level beating on your last day in prison. But both of you could have walked out of prison with only a mild beating on your last morning if you had not acted in your own self-interest. In any Prisoner's Dilemma game the paradox is always the same: by rationally acting in your own self-interest, you harm not only another individual, but yourself as well.

It turns out that humans—and other species as well—face Prisoner's Dilemma-like situations constantly. As I said earlier, these situations obviously need not involve prisoners. In fact, the only real requirement for a situation to qualify as a Prisoner's Dilemma is that the given payoff matrix have the follow-ing form: For each player, the benefit from "cheating" (in my example, not accusing Bruiser when your cell mate did), must be greater than the benefit from "cooperation" (accusing Bruiser when your cell mate did as well), which must in turn be greater than the benefit from "mutual noncooperation" or "defection" (not accusing Bruiser when your cell mate did not accuse Bruiser), which must finally be greater than the benefit of "being cheated" or "getting the sucker's payoff" (accusing Bruiser when your cell mate did not).[4] The beauty of the Prisoner's Dilemma—and the reason that it enjoys such wide applicability in domains like mathematics, economics, evolutionary psychol-ogy, and so forth—is that it helps to clarify the relationship among truth, rationality, self-interest, and altruism.

Consider, once again, your reasoning in the above example. For the sake of clarity I have diagrammed the payoff matrix in table 1. Your choices are to be found in the two columns of the matrix, while your cell mate's choices are to be found in the two rows. The payoffs to you from the four combinations of these choices are presented in italics, while the payoffs to your cell mate from the four combinations of these choices are presented in boldface.

		You	
		Accuse Bruiser	Do Not Accuse Bruiser
Your **cell-mate**	Accuse Bruiser	*Mild beating* **Mild beating**	*No beating* **Severe beating**
	Do Not Accuse Bruiser	*Severe beating* **No beating**	*Medium beating* **Medium beating**

As you examine the matrix, notice that it pays both you and your cell mate simply to tell the truth. By doing so, both of you end up with the best possible joint outcome. Indeed, if either of you lie, both of you are hurt. Thus in some perhaps "metaphysical" sense, truth and self-interest are related in this Prisoner's Dilemma situation. The rational, self-interested thing to do, it would seem, is to tell the truth. Hence we might be tempted to say that, *in general*, the Prisoner's Dilemma is not really a "dilemma" after all, since the seeming dilemma is caused not by rational, self-interested action (i.e., telling the truth) but, rather, by irrational action (i.e., lying).

Unfortunately, this is wrong. It turns out that truth need not *necessarily* be the best policy in *all* Prisoner's Dilemma situations. Suppose that you and an accomplice are arrested and charged with conspiracy to commit murder. Assume that you are both guilty, and that you are each taken to separate holding cells. Assume, finally, that each of you knows that the evidence against both of you is quite flimsy. Presently, the district attorney in the case enters your cell and presents you with the following deal, which he tells you he has *also* presented to your accomplice: If you confess to the conspiracy and implicate your accomplice, while your accomplice remains silent, then you will receive parole and your accomplice will get ten years in prison. If, on the other hand, you remain silent, while your accomplice confesses and implicates you, your accomplice will receive parole and you will get ten years in prison. If both of you confess and implicate each other, both of you will get five years in prison. But, lastly, if both of you remain silent, both of you will be held in prison as material witnesses for one year and then released. What do you do?

In this Prisoner's Dilemma situation, telling the truth, the whole truth, and nothing but the truth, actually hurts both you and your accomplice. Instead, it would be best for both of you if both of you lied as thoroughly as possible—by not confessing yourself and by not implicating your accomplice. In this case, then, with respect to you and your accomplice, truth is not in your *mutual* self-interest.

Doubtless many individuals, especially those with a philosophical bent, will rebel against the thought that truth is not always in everyone's self-interest. Indeed, the idea that truth is an indivisible good, and that what is good for one is necessarily good for all, is at the heart of Platonism (and, I would argue, most major religions). This may explain why R. E. Allen—perhaps the best modern English-language translator of Plato—comes down so hard on the whole idea of the Prisoner's Dilemma. Allen argues that even in the case where you and an accomplice have conspired to commit murder and have been justly arrested, it is still in your self-interest to confess and implicate your accomplice, since doing so guarantees just punishment for both of you, and just punishment is always in a wicked person's self-interest, whether he realizes it or not.[5]

The problem with this "solution" to the Prisoner's Dilemma is that it unites truth, rationality, and self-interest simply by definitional fiat. To be sure, if you believe in the *ultimate* unity of your self-interest with the self-interest of every other human (or even every other living being), then you could never face a Prisoner's Dilemma situation. In particular, you could never be in a position in which your apparent self-interest and the apparent self-interest of another conflicted in a way that could not be resolved rationally, simply by getting to the "truth" of the matter. But notice that if you do believe in such an expansive unity of self-interests, you could never be an altruist, since *genuine* altruism involves *sacrificing* your own self-interest for the interest of another. Where there is no genuine conflict of self-interests, there can be no genuine sacrifice by one individual of his or her own interest for the interest of another. The bizarrely paradoxical result would seem to be that the Prisoner's Dilemma—by positing a genuine conflict of self-interests—opens up the possibility for genuine altruism, but then forecloses the possibility that such altruism could be the result of rational, self-interested action.

Does this doom humankind to a world devoid of altruism of any kind? Is "survival of the fittest" really the best we can hope for as a species? Is this what the Prisoner's Dilemma means for humanity? Perhaps not. To see why, recall that in the original Prisoner's Dilemma situation I outlined above Bruiser (that sly guard) replaced your long-time cell mate with another prisoner, the very day before you were to be released. Your new cell mate was on the same cellblock, and hence knew the "system," but had no acquaintance with you. That meant that you and your new cell mate, who was also scheduled for release with you the next day, "played" what is called a *one-time-only* Prisoner's Dilemma game. Such games are different from *iterated* (or repeated) Prisoner's Dilemma games, and that difference literally makes all the difference in the world. The reason is obvious, and it is also essentially the reason that Bruiser replaced your old cell mate. Bruiser reasoned that you and your old cell mate might well have built up a high level of trust over the years you had been together in the cell.

What is *trust*, really? It is the feeling a person has that when he or she plays a Prisoner's Dilemma game with another individual that other individual will *cooperate to both of their mutual benefits*. In my initial example, mutual cooperation would have meant that both prisoners would have accused Bruiser of mistreatment. That would have resulted in *both* prisoners receiving the highest "benefit" (i.e., a mild beating) that was possible given the particular Prisoner's Dilemma payoff matrix. Obviously, such trust between two individuals cannot emerge unless there are at least two Prisoner's Dilemma games played together—that is, unless two players are playing an iterated Prisoner's Dilemma game.

Actually, there is a further requirement that I must also discuss. In order for there to be the possibility that trust might emerge from an iterated Prisoner's Dilemma game, neither player can know when the game will end. If a player knew what was to be the last move of the game, he would obviously not cooperate on that move. If the other player was trusting and *did* cooperate on that last move, then the noncooperating player would receive the highest payoff possible (the payoff given to the "cheater") while the cooperating player would receive the lowest payoff possible (the "sucker's payoff"). Of course, if *both* players knew what was to be the last move of the game, neither player would cooperate on that move. Hence on that last move the iterated Prisoner's Dilemma game would become a one-time-only game with the outcome being suboptimal for both players. But it is even worse than that. If each player knew what was to be the last move of the game, and knew that the other player knew this as well, then both players would "defect" (i.e., not cooperate) on the *second-to-last* move of the game, and so forth backwards to the move they were just about to play. Thus an iterated Prisoner's Dilemma game in which both players knew when the last move would occur would become a one-time-only Prisoner's Dilemma game.

Notice that because of this it may *not* have been necessary for Bruiser to remove your old cell mate from your cell. Assuming you would not interact with your old cell mate once you left prison, you may well have reasoned that the meeting you had in the warden's office *was* the last move of the two-year iterated Prisoner's Dilemma game you had been playing with your old cell mate. But Bruiser was not willing to take that chance, so he hedged his bet by placing in your cell an inmate you did not know.

If most inmates *do* know when their sentence is up and are *not* likely to interact with fellow inmates once they leave prison, it *may* be uniquely difficult for trust to emerge in prisons. But surely there would still be plenty of opportunities in prison for two cell mates to be placed together in Prisoner's Dilemma-like scenarios on repeated occasions. Think about it. A particularly ill-tempered prisoner is harassing your cell mate. Do you come to your cell mate's defense, even though you might get a black eye? Would your cell mate do the same for you if the situation were reversed? Your cell mate has missed supper. Do you share some of the canned tuna you have with him, even though that means you will have a little less? Will he do the same with his canned Spam when you miss supper? Your cell mate says he is feeling ill. Do you help him out by filling in for him on laundry detail, even though that means you may need to work a little longer? Not all of these particular scenarios need be Prisoner's Dilemma games, *in the strictest sense*. But they would all surely afford the opportunity for trust to develop, *or not develop*, between these prisoners.

Of course, there is trust, and there is trust. Put another way: there is *blind* trust and there is a, let us say, more *measured* type of trust. I am reminded here of former President Ronald Reagan's famous "maxim"—developed in the latter part of the 1980s—regarding the approach that the United States should take toward the former Soviet Union with respect to arms control agreements: *trust but verify*. This seems like an appropriate strategy for two nations that had, over the course of several decades, been playing an iterated Prisoner's Dilemma game, but never all that successfully. At least Reagan's maxim avoided the two poorest strategies in any iterated Prisoner's Dilemma game: the strategy of being blindly trusting and the strategy of being blindly distrusting.

To see why these strategies are so poor, suppose you are involved in an iterated Prisoner's Dilemma game. On each move you blindly trust that the other player will act cooperatively. After several moves—that is, after several choices involving different scenarios like accusing Bruiser, sharing food, helping at the laundry, and so forth—the other player will realize that you are (to put it bluntly) a sucker. You will then be continually taken advantage of by this player. On the other hand, suppose you always blindly distrust the other player in the game. In such a case, you would always make the "mutual noncooperation" choice for each move. This is exactly equivalent to playing an iterated Prisoner's Dilemma game *as if* on each move you were playing a one-time-only Prisoner's Dilemma game. The result would be that you would never get beyond the paradox associated with the one-time-only game. Surely, between blind trust and blind distrust there must be a good strategy for playing an iterated Prisoner's Dilemma game. And if there is a good strategy, there is a best strategy. What might that strategy be?

In the early 1980s the political scientist Robert Axelrod set out to answer this question. He developed a computer program that simulated a *formal* iterated Prisoner's Dilemma game. He then invited game theorists, and anyone else who cared to do so, to submit "strategies" for playing this game. Each strategy was fed into the computer and played against every other strategy. One emerged victorious. Later work has shown that this is, in fact, the best strategy to adopt when playing an iterated Prisoner's Dilemma game of any type. "Best" has a very specific meaning here. This strategy is best in the sense that it provides the individual playing it with the maximum benefit he could get from any strategy, and it provides the maximum *overall* benefit that could accrue to *everyone* in the game if everyone in the game adopted the strategy. Finally, the strategy is evolutionarily successful over all other strategies in the sense that individuals playing the strategy, and then "reproducing" in proportion to the "benefits" they derived from the strategy, will tend eventually to outreproduce a population playing any other strategy or combination of strategies.[6]

What is this amazingly successful strategy? It has come to be called TIT FOR TAT. In an iterated Prisoner's Dilemma game an individual who adopted a TIT FOR TAT strategy would always cooperate with the other player on the first move and then do on each subsequent move exactly what the other player did on the previous move. Thus a person playing this strategy would, for example, begin the game in the spirit of cooperation by, let us say, accusing Bruiser. If the other player also cooperated by accusing Bruiser, the person playing the TIT FOR TAT strategy would then cooperate on the next move by, for example, sharing food. If the other player, however, then failed to cooperate by sharing, the person playing TIT FOR TAT would then *not* cooperate on his next move. If the noncooperating player then decided to resume cooperating on some future move, the person playing the TIT FOR TAT strategy would then cooperate on the move immediately following that one. The strategy is that simple.

Now if TIT FOR TAT were a "person" instead of a "strategy," what kind of person would it be? First, it would be a naïvely optimistic person. Upon meeting another it would always think the best of him or her. It would always seek to cooperate. Trivers reports on an endearing study which found that very young children are more likely to share a valuable toy with a stranger than a friend. Interviewed after the study, some children innocently explained that they were just trying to make new friends.[7] But, second, if TIT FOR TAT were a person it would also be just and rational. If it were taken advantage of, it would retaliate immediately, but in a measured way. It would do unto others as was done unto it. Finally, TIT FOR TAT would be a forgiving person. If you took advantage of it, but then attempted to cooperate, it would forgive you your past sin and reciprocate your cooperation.

That last point brings us back to Robert Trivers's initial article on reciprocal altruism. If reciprocal altruism is what occurs when helpful people help other helpful people, reciprocal altruists may be characterized as players in an iterated Prisoner's Dilemma game who adopt a strategy of TIT FOR TAT. And like TIT FOR TAT, reciprocal altruism is a kind of altruism *that goes beyond kin*. This is an absolutely critical point. As Trivers emphasizes, in reciprocal altruism "it is the *exchange* that favors such altruism, not the fact that the allele [i.e., specific copy of a gene—EMG] in question sometimes or often directly benefits its duplicate in another organism."[8] It is precisely this ability to get beyond biology that gives reciprocal altruism its power. Indeed, it is precisely this ability that allows for the possibility—and in many cases the actuality—of reciprocal altruism even between members of different species.

But there are some fairly major assumptions built into Trivers's model of reciprocal altruism. One is that individuals in the species live long enough to make reciprocity itself possible. Obviously, human beings do. Another is that there exists a "low dispersal rate" between individuals in a given locality, and that there be fairly frequent opportunity for interaction between them. It seems highly likely that this was the case in the environment of our hominid ancestors. Recall the discussion of Pleistocene life in chapter seven. Yet another assumption involves the nonexistence of an extremely strong "dominance hierarchy" in the group. Obviously, if one individual in the group is so strong as to be able to command from others all or nearly all of their resources at all times, this would make the possibility of reciprocal altruism extremely difficult because it would make *any* exchange—other than the giving of all resources to the dominant individual—difficult. It is possible that ancestral humans lived in such an environment. But it is probably not likely. Indeed, one of the conditions of the ancestral environment that may have made strong dominance hierarchies rare is that they would have been difficult to enforce, given the very *possibility* of reciprocal altruism in mutual combat. Consider that while one male may be stronger than another in a given community, and perhaps even stronger than any other single male individual in the community, it is unlikely that any male would be stronger than a group of almost *any* three or four other males in the community.[9]

It seems likely, then, that reciprocal altruism between ancestral humans would have been quite possible if the opportunity for it ever arose, and it seems likely that there were an abundance of such opportunities. Taking advantage of even a small number would likely have created other opportunities. If only a few of those opportunities were seized, this would have created still more opportunities, and so on, in what I earlier described—in the context of sexual selection—as a positive feedback loop. One could persuasively argue that the human mind evolved as quickly and complexly as it has largely to take advantage *as effectively as possible* of the opportunities provided by reciprocal altruism.[10]

Consider some of the obvious opportunities for reciprocal altruism that would have existed in the ancestral environment, and note simultaneously how this list sounds much like the list I drew up regarding situations that might have made it possible for trust to emerge in a prison environment. One opportunity for reciprocal altruism in the ancestral environment has already been mentioned: mutual aid in combat. Suppose I aid you when you are being harassed by another individual. That is a benefit to you but a cost to me. Suppose you do the same for me, sometime later when I am being harassed. If we mutually cooperate in an environment that contains the possibility of harassment, we would be selected for over noncooperators. Next consider food sharing. As I said in chapter seven, there were obviously no food storage devices in the ancestral environment. This would create a clear opportunity for reciprocal altruism. If I happened to have a very successful day of hunting, I might share some food with you. If you are successful another day, you might share with me. Of course, one needs to be careful on the hunt. Injuries occur. Such injuries are yet *another* opportunity for reciprocal altruism. I lean on you today, you lean on me tomorrow. This list is almost endless. Trivers adds many other examples: helping in times of danger (with warning calls, for example); helping in the care of children; sharing implements; and sharing knowledge (undoubtedly the most frequent opportunity).[11]

The society that emerges from the practice of reciprocal altruism is precisely the type of society that emerges when a TIT FOR TAT strategy is adopted in an iterated Prisoner's Dilemma game. This society does seem strikingly like Walnut Grove, the fictional town depicted in "Little House on the Prairie." It is a society in which helpful people help other helpful people. In such a society, any individual's gains could be greater than the gains he could achieve alone, and no individual need necessarily suffer a loss because of the gains of anyone else. This is what economists call a non–zero-sum environment. The fact that reciprocal altruism can produce such an environment makes it an extremely powerful and progressive force in human evolution.

But if, as it has been said, eternal vigilance is the price of freedom, such vigilance is also the price of reciprocal altruism. For just as there are those who would threaten freedom, there are also those who would threaten reciprocal altruism. In the nineteenth-century town of Walnut Grove such threats may often have come in the form of confidence men. They would solicit your cooperation without cooperating in return. They would cheat you out of house and home. What is a cheater? Simply someone who takes a benefit without reciprocating. Thus perhaps the *greatest* threat to a society of reciprocal altruists comes from the presence of cheaters in their midst. If reciprocal altruism is so powerful, and so important to the human species, we might expect that evolution has equipped us with a special "defense mechanism" against cheaters. Has evolution done so? And if so, what is this mechanism?

Recent research by evolutionary psychologists has come up with an amazing answer to these questions. The answer has profound implications for the study of logic and reasoning. To get a sense of this research, I invite the reader to answer the two questions that appear below.[12] Suppose that you are the chairperson of a communication studies department at a university. Your dean is doing another one of his esoteric studies on student course enrollments and demographic groups. You neither know nor care what the study is designed to show, but for the purposes of the study, students in your department must be sorted based on the following rule:

> Rule 1: If a student is currently enrolled in "Public Relations" then he or she must be in Group 3.

Your student assistant has been sorting student data all day. But because he is a sophomore, you suspect he may have become confused and not followed the above rule. Below are four cards. Each card corresponds to one student. One side of the card indicates the one and only communication studies course for which the student is currently enrolled and the other side indicates the one group the particular student is in. Here is your first question: *Which one(s), if any, of these cards must you turn over to be absolutely certain that Rule 1 has been followed?*

Currently enrolled in "Public Relations"	Currently enrolled in "Media Analysis"	Group 3	Group 5

After answering that question, consider another very similar situation. Suppose that you are the chairperson of a communication studies department at a university. As it happens, everyone in the university wants to take a course in your department called "Public Relations," because experience has shown that students who take this course can easily get high-paying jobs. In order to limit the number of students in "Public Relations," and to ensure that the students who are in the course are serious and deserving, you impose a *requirement* for enrollment in the course, in the form of the following rule:

> Rule 2: If a student is currently enrolled in "Public Relations" then he or she must be a senior.

You suspect that some students may have cheated and gotten into the course without following the above rule. Below are four cards. Each card corresponds to one student. One side of the card indicates the one and only communication studies course for which the student is currently enrolled and the other side indicates whether the student is a freshman, sophomore, junior, or senior. Here is your second question: *Which one(s), if any, of these cards must you turn over to be absolutely certain that Rule 2 has been followed?*

Currently enrolled in "Public Relations"	Currently enrolled in "Media Analysis"	Senior	Junior

After answering these questions, you may notice that they both have exactly the same logical form—*If P then Q*—where *P* corresponds to *currently enrolled in "Public Relations"* and *Q* corresponds *either* to *Group 3* or *Senior*. The negation of an if-then statement of this form is: *P* and *not Q*. Hence, for *both* questions above, the correct answer is that you would need to turn over only the first card and the last card, because only on those cards could you possibly encounter a case of *P* and *not Q* on the same card.

You might think that individuals would get the correct answer to *each* of these questions as often as they got the incorrect answers since both questions have exactly the same form. Only their content is different. But, strikingly, this does not seem to be the case. In fact, individuals do dramatically better in correctly answering the second question, by a ratio of about 3 to 1.[13] Just because both of these questions take the same form, the observed discrepancy *must* therefore have something to do with the content of each question. Look again at the second question. In that question *P* can be understood as a "benefit." We know that students can get high-paying jobs if they take "Public Relations." Similarly, in the second question *Q* can be understood as a "cost." Even if being a senior is not a "cost" in the negative sense of the term, it is at least a requirement. Now, a person who takes a benefit without paying the requisite cost or meeting the same requirement as everyone else (*P* and *not Q*) is a cheater. Everyone knows that no one likes a cheater. Now we may know something about why cheaters are easy to spot.

Experimental research by the evolutionary psychologists Leda Cosmides and John Tooby strongly indicates that individuals reason dramatically better when they are reasoning about a situation that in some way implicates a form of social exchange *that is potentially open to cheating* than when they are reasoning about a simple descriptive rule (*If a student is currently enrolled in "Public Relations" then he or she must be in Group 3*) or even about a causal relationship (*If the room has windows then students will be distracted*). Cosmides and Tooby argue on the basis of this evidence that all humans come equipped with a mental module specifically designed for detecting cheaters.[14] Although Cosmides and Tooby do not use this analogy, one might think of our cheater detector module in the same way we think of our face recognition module. In both cases the modules are specialized, not general. A face recognition module is good at recognizing the shapes and features of human faces only. It is not good at recognizing shapes and features of objects in general. Similarly, a cheater detector module is good at "reasoning" about situations that might involve cheating. It is not good at reasoning in general.

Research also shows that when the rules of formal logic differ from the "rules" used by our cheater detectors we are *better* able to detect cheaters by using the cheater detector than we would be by using the rules of formal logic. Cosmides and Tooby argue, correctly I think, that the following two rules are logically different, but equivalent from the perspective of a social exchange.

Rule 3: If you give me your watch, I'll give you $20.

Rule 4: If I give you $20, you give me your watch.[15]

Notice that the formulation of Rule 3 is identical to the formulation of the above Rules 1 and 2. Thus in Rule 3 P—always the first clause in the conditional statement—corresponds to *you give me your watch* while Q—always the second clause in the conditional statement—corresponds to *I give you $20*. Notice also that in Rule 3 P is the benefit (to me) and Q is the cost (to me) in the exchange.

But for Rule 4 P corresponds to *I give you $20* while Q corresponds to *you give me your watch*. Thus for Rule 4 P is the cost (to me) and Q is the benefit (to me) in the social exchange. But, as Cosmides and Tooby write, "No matter how the contract is expressed, I will have cheated you if I accept your watch but do not offer you the $20, that is, if I accept a benefit from you without paying the required cost."[16]

Now suppose you show two groups of individuals the following sets of cards.

You gave me your watch	You did not give me your watch	I gave you $20	I did not give you $20

Suppose further that you gave one group Rule 3 and asked that group which cards would need to be turned over to detect violators of that rule, while you gave a second group Rule 4 and asked that group which cards would need to be turned over to detect violators of that rule. If we approach social exchange situations that implicate the possibility of cheating *logically*, we would expect that the first group would do substantially better at the assigned task than the second group. This is because the first group was working from a rule that was logically equivalent to the conditional *If P then Q*, and could thus be negated by *P* and *not Q*. But if those in the second group used the negation *P* and *not Q* as applied to *their "switched" formulation of the rule*, they would turn over the third card *P* and the second card *not Q*—exactly the wrong two cards. Remarkably, both groups do equally well at detecting cheaters and, again, much better than they do at detecting violations of simple descriptive rules.[17] It seems, then, that when we "reason" about social exchange situations that might involve cheating we turn "off" our logical reasoning module and turn "on" our cheater detection module.

Further, there is evidence suggesting that the content specific nature of the cheater detector module is extremely fine tuned. It appears that the module gets turned on only when we reason about a social exchange situation that involves the possibility of cheating, but also, that the module gets turned on in these situations *even if* we do not understand the cultural context of the social exchange. For example, most Americans would doubtless understand the following statement

Rule 5: If you vote in a federal election then you must be a U.S. citizen

as a type of social exchange situation open to the possibility of cheating. An individual may try to vote without being a citizen. Thus if you were to show Americans the following four cards

Voted in a federal election	Did not vote in a federal election	Is a citizen	Is not a citizen

you probably would not be surprised if they were good at detecting violators of this rule. You might suspect that their success came not from the use of any cheater detection module, nor even from the use of any reasoning process, but rather from the fact that Americans are simply familiar with this aspect of their culture. But this appears *not* to be the case. When subjects were given a simple *descriptive* rule with which they could be expected to be familiar—such as, *If one goes to Boston, one takes the subway*—they were nowhere near as good at detecting violations of this culturally familiar rule as they were at detecting violators of a culturally familiar rule that implicated the possibility of cheating in a social exchange situation.

But what clinches this point is an examination of how well people do in detecting violators of rules when they have absolutely no familiarity with the cultural context in which the rule is embedded. Two groups of people were given the following rule:

Rule 6: If a man eats cassava root then he must have a tattoo on his face.

They were then shown the following cards:

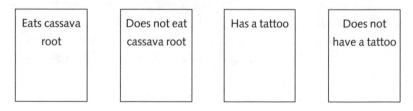

Eats cassava root	Does not eat cassava root	Has a tattoo	Does not have a tattoo

Notice, first, that this rule does not correspond to any cultural practice with which any subject would likely be familiar, because it was simply made up by the researchers. Notice also that this rule need not *necessarily* implicate a social exchange situation involving the possibility of cheating. In fact, one group was told that in the particular culture from which this statement was drawn having a tattoo on one's face meant that one was married, and all married men just happened to live on the side of the island on which only cassava root grows. This explanation makes the above rule a simple descriptive rule, similar to the rule that if one eats sauerkraut then he must be German. The other group, however, was told that cassava root is a highly desirable food that not everyone is allowed to eat. It was further explained that having a tattoo signifies that one is permitted to eat it. Remarkably, the second group did dramatically better, by a margin of three to one, in detecting violators of the rule than the first group, even though the rule and the cards were exactly the same for both groups.[18] The almost inescapable conclusion is that the second group had their cheater detectors activated.[19] Also, in general, individuals do better at detecting violators of culturally *unfamiliar* rules that implicate the possibility of cheating in a social exchange situation than they do at detecting violations of culturally *familiar* rules that are merely descriptive but do not implicate the possibility of cheating.[20]

Finally, consider this. If you were designing a mental module to be used by ancestral humans for whom social exchange was a vital part of life, and if *efficiency* were a critical concern—remember any module eats metabolic energy and takes up brain space—what would be the minimum requirements for this module? Obviously, you would want it to be good at detecting cheaters. But would you necessary want it to be good at detecting altruists? Probably not, since altruists pose no threat to society. And, indeed, it appears that while we do have a mental module for detecting cheaters, neither that module, nor any other module we may have, works very well at detecting altruists. Subjects were given many of the same rules and cards I have been discussing above, and asked which cards they would need to turn over to determine who had "violated" the rule by being altruistic—that is, by paying a cost but not taking a benefit. Subjects did no better at this task than they did at determining violations of simple descriptive rules.[21]

All of this evidence and more seems to suggest strongly that we do have a specific mental module for cheater detection, and that this module is not a spandrel or by-product of our general ability to reason. It is a hardwired, "dedicated" module designed to focus specifically on one set of "inputs" (the possibility of cheating in a social exchange situation) and return one set of "outputs" (the benefit-cost structures that are necessary to evaluate whether one has been cheated). Perhaps the most significant implication of this research is that the *process* of human reasoning—a process that is still thought to be so insensitive to content that the premises of arguments can be represented in the formal logic as merely letters like *P* or *Q*—is itself different depending upon the *content* of those *P*s and *Q*s. If this is true, we may need to rethink, for example, the way in which we administer intelligence tests—specifically tests that purport to measure an individual's skill at inferential reasoning. From now on, we may need to specify which inferential reasoning skills—for example, ones about social exchange or ones about descriptions of the world—that we are attempting to measure, and we may need to formulate the content of the questions accordingly. Cosmides and Tooby also note that if their findings hold up, we may be justified in looking for different reasoning processes in other areas of life, including the evaluation of threats, the benefits associated with joining certain "coalitions" of other humans, and mate choice.[22] Finally, the existence of a cheater detector fits *so perfectly* with what we would reverse engineer as a requirement for the practice of reciprocal altruism that we seem to have extremely strong evidence for viewing reciprocal altruism as a vital aspect of the life of ancestral humans. Might we go further and draw the conclusion that reciprocal altruism was the first—and perhaps for that reason still the strongest—ethical and political "philosophy" devised by humans?

I think we *should* draw this conclusion. Indeed, I would argue that reciprocal altruism is the ethical and political philosophy underlying all other ethical and political philosophies precisely because it best embodies our innate sense of justice and fair play. What, after all, is the famous Golden Rule—"Do unto others as you would have them do unto you"—if not a colloquial expression of reciprocal altruism? If you want a less colloquial, and decidedly more philosophical, formulation of exactly that rule you can turn to Immanuel Kant, who offers us the "categorical imperative." The clearest statement of that ethical command can be put as follows: Act *only* so that the maxim embodied in your particular action could be a universal rule for everyone.[23] This command would prohibit lying and stealing, since if everyone lied and stole then society could not function. It would also prohibit cheating for the same reason. If you couple either of these rules with the injunction to "forgive and forget," you have not only reciprocal altruism colloquially expressed, but also all the requirements of a TIT FOR TAT strategy in an iterated Prisoner's Dilemma game.

Of course there are ethical systems that seem to ask more of their followers than simple reciprocal altruism. Christianity is often interpreted as such a system. Thus in St. Matthew's description of the Sermon on the Mount, Christ offers his followers a new covenant based on what appear to be blindly altruistic commands:

[W]hosoever shall smite thee on thy right cheek, turn to him the other also.

And if any man will sue thee at the law, and take away thy coat, let him have thy cloke also.

And whosoever shall compel thee to go a mile, go with him twain. (Matthew 5: 39–41).

These commands seem difficult, if not impossible, for any viable human *society* to embody in its ethical or political philosophy. Further, I think these commands would be difficult, if not impossible, for any reasonably normal and effectual individual to live by on a daily basis. In fact, whatever one thinks of the overall value of Christian ethics (and clearly it has much value), I think Nietzsche was right on target in pointing out that the injunctions in the Sermon on the Mount would be psychologically untenable if they were not also coupled with the promise that there is a "just" God who will balance the scales of reward and punishment in the hereafter.[24] The last shall be first, we are told. In this regard it is striking that only four verses after the last verse quoted above, St. Matthew has Christ say: "For if you love [only] them which love you, what *reward* have ye?" (Matthew 5: 46). In the eschatological context of the Sermon on the Mount, the clear implication here would seem to be that only those who follow the altruistic commands of the new covenant will receive the very great benefits to be had in the afterlife. To be sure, this still is not reciprocal altruism. But neither is it a system in which cheaters get away with their evil deeds. They are punished—if not by man, then by God.

The notion of *eventual* punishment seems critical not only to the psychological tenability of any such *seemingly* blindly altruistic system, but also to the development of our senses of guilt and conscience. One could easily reverse engineer our sense of guilt as just the type of mechanism that would likely develop in a species for which social exchange mattered so greatly to everyday life, for which the temptation to cheat when engaging in such exchanges was therefore correspondingly great, but also for which the harm attached to cheating—i.e., being found out and hence being distrusted in the future—was itself correspondingly great. A conscience and a sense of guilt would keep in check one's natural (rational?) tendency to cheat. It does not take an evolutionary psychologist to see that this sense of conscience might easily be projected onto an omniscient being who could infallibly detect cheating.

There is, then, compelling evidence that at least some of the major philosophical and religious systems in the Western world are grounded on the ethical principle of reciprocal altruism, or at least that when such systems seem to be opposed to this principle—as *may* be the case in Christianity— these systems tend to develop mechanisms (i.e., conscience and a belief in an omniscient and *just* God) that bring their own workings back in line with what one would expect from a system based on reciprocal altruism. But the best evidence that reciprocal altruism forms something of a human ur-ethics may come from the observed fact that humans seem to spontaneously live by the principle embedded in this system. I think that supermarkets provide some evidence of this claim.

How often have you been in a supermarket and proceeded to a checkout line clearly labeled TEN ITEMS OR LESS only to find that the person three customers in front of you is in line with twelve items?[25] And it is always twelve or thirteen items that one finds in the basket of this potential cheater, not forty or fifty. Almost everyone, I would bet, finds this distressing, even if he or she does not *say* anything. But there may be one affected individual who does remind the cheater that the sign stipulates ten items or fewer. The cheater may respond verbally, or more often with rolled eyes and a sigh that says: "How petty can you be; it would take only a fraction of a second for the checkout clerk to ring up my extra two items." But the accuser may also respond verbally, or more often with a moralistic stare that says: "The point is not the time involved in ringing up the extra two items; it is rather the principle involved. If everyone behaved as you do the system would break down."

I think that for most people it really *is* the principle involved. This helps to explain, as Trivers notes, why moralistic denunciations of cheaters often seem "out of all proportion to the offenses committed."[26] An individual who takes *only* one or two more than the allowed number of items to a checkout line is hardly harming anyone else to any *significant* extent. But the denunciation of that individual's behavior may be most effective if it is strong and unmistakable, thus sending the signal that the system of reciprocal altruism must be respected.

Of course, it is not too difficult to *detect* a potential cheater in a checkout line reserved for those with ten items or fewer. One simply counts the number of items in people's baskets. Nor is the denunciation of the twelve-item cheater an explicit example of one's cheater detector at work. Still, the fact is that we *do* seem to pay very close attention to the potential for cheating in a social exchange situation, rather than to the *general* aspects of the situation. Ask yourself how many times you have been in an express checkout line and counted, not the number of items in the baskets of persons in front of you (to see who might be a cheater), but rather the number of items in the baskets of the individuals in the "regular" checkout lines (to see who might be an altruist by *not* utilizing the express checkout line when he or she was entitled to do so). My point is that our focus on potential cheaters, whether their actions be subtle or not so subtle, strongly suggests the existence of a cheater detector module that is attuned to very specific "inputs" from the environment, and that this module makes reciprocal altruism both possible and likely.[27]

Notice, by the way, how critically different would be your reaction if in my little example the sign in question had read FOR WHITES ONLY, and if the "potential cheater" in line had happened to be a black individual. In such a situation, our sense of moral outrage should (and I think would) be directed toward anyone who attempted to "denounce" the black individual, and indeed toward whoever put up the sign. It would perhaps be too much to suggest that reciprocal altruism *guarantees* that humans will spontaneously act to secure equal rights for all individuals in a given community. But it may not be too much to suggest that reciprocal altruism makes talk of equal rights easier precisely because reciprocal altruism seems to foreground *basic concepts of fairness*. At any rate, it is undeniable that the most persuasive arguments in American society today are those that appeal to fundamental fairness as their basic value. Everyone realizes this, and so everyone structures his or her public arguments accordingly.

Thus you are unlikely to see a striking union worker carrying a picket sign that explicitly announces a demand for more money, even if that demand were made on behalf of that worker's impoverished children. Rather, the union worker's sign will probably imply—and the worker himself or herself will surely believe—that the real issue is not about money, but instead about fair treatment. The worker is giving something (his or her labor) to the employer, but the employer is trying to cheat the worker by not giving something (a salary) *of equal value* in return.

A similar rhetorical approach is used by "liberal" politicians who would have us increase taxes on the wealthy so that they are made to pay "their fair share." The rhetorical appeal here is never made simply in the name of "soaking the rich." To be sure, there is quite a bit of soaking that these politicians and their constituencies might like to see. But the appeal is always made in the name of fairness. Naturally, this appeal can be all the more successful when the wealthy are seen as not really deserving the money that they possess. Thus it is not surprising that within certain discursive circles capital gains and profits from investments are referred to as "*unearned* income."

Of course "conservative" politicians are equally adroit at using a political rhetoric that foregrounds fundamental fairness and warns us to be on the lookout for cheaters. Consider the fairly recent history of welfare reform in America. During the 1980s President Reagan went a long way in laying the groundwork for dismantling the federal welfare bureaucracy, and curtailing the overall amount of welfare payments to individuals, by insisting that large numbers of individuals were abusing the welfare system. "Welfare queens," as they came to be known, were supposedly everywhere, driving Cadillacs and wearing expensive clothes.[28] In the 1990s Bill Clinton then went on to complete the welfare revolution by restructuring the system along lines that were not all that different from those laid down by Reagan. Significantly, Clinton did this in part by relying on a similar, though perhaps a gentler, version of Reagan's arguments. Recall Clinton's pledge in his 1992 acceptance speech at the Democratic National Convention to "end welfare as we know it," and his promise to say to those on welfare: "You will have, and you deserve, the opportunity through training and education, through child care and medical coverage, to liberate yourself. But then, when you can, you must work, because welfare should be a second chance, not a way of life."[29]

From a rhetorical perspective that is also informed by evolutionary psychology, the public policy debate surrounding welfare unfolded in ways that seem quite consistent with the rules governing reciprocal altruism. Notice that both Clinton and Reagan framed the welfare debate as one in which a benefit was being altruistically offered to members of the community. It is significant that Clinton does *not* say that individuals deserve welfare. They only deserve an opportunity to better themselves. Welfare is a type of governmental "charity" given to individuals so that they can help themselves, and then, in the future, become taxpayers and thus help others, who also want to help themselves. This is reciprocal altruism any way you look at it.

Notice also that both Clinton and Reagan saw that what disturbed most Americans was not the existence of welfare as such but rather the potential for cheating the system, and, more importantly, the inability of individual Americans *directly to detect such cheating*. Huge welfare bureaucracies may be good at taking advantage of economies of scale when delivering their "product," but they wildly set off our cheater detectors. Yet, because of their very size, welfare bureaucracies prevent individual taxpayers from effectively monitoring the system. This helps to explain an aspect of the welfare debate that bedeviled Ted Kennedy liberals, and also that probably caused them to think badly of their fellow citizens. Throughout the 1980s, and especially in the early 1990s, liberals were saying, quite correctly, that the whole welfare debate was grossly out of proportion to the amount of money that welfare payments themselves represented as a percentage of the overall federal budget. Liberals wondered how average Americans could be so exercised over so trivial a percentage of the federal budget, especially when other areas of the budget—defense spending, for example—went seemingly unscrutinized. Liberals concluded that Americans must be greedy and selfish.

But this conclusion was simply wrong, for it failed to take account of precisely how our cheater-detection mechanism works. Welfare is a particularly salient example of social exchange. Thus it immediately sets off our cheater detectors. Even the possibility of cheating then becomes significant, precisely because if cheating takes place it tends to spread quickly. This, at any rate, is a type of thinking that would be quite consistent with what we know about human reasoning with respect to reciprocal altruism. Hence it may not be that the average American is greedy and selfish. It may rather be that people everywhere have a natural impulse to enforce *responsibility*, especially with respect to social exchange situations. This impulse was clearly reflected in the welfare debates of the 1990s, including in the title of the very bill that was being debated. Although it is sometimes called the "welfare reform" act for short, we should not forget that on 22 August 1996 President Clinton signed what is formally known as the "Personal Responsibility and Work Opportunity Reconciliation Act."

With that said, the examples I have used so far include some fairly quantifiable senses of gain and loss, from the perhaps trivial amount of time lost in a supermarket checkout line to the more substantial gains and losses related to wages, taxes, and welfare payments. But in truth, there is almost no area of human interaction in which the concept of fundamental fairness, the calculus of reciprocal altruism, and the looming danger of being cheated is not relevant.

Consider the issue of affirmative action and racial preferences, specifically as they have developed in America over the course of the last forty or so years. To its supporters, affirmative action is a way of ensuring more nearly equal representation among the races in all areas of American life, but particularly in employment and higher education. To its opponents, affirmative action is nothing more than racial preferences for less qualified, "underrepresented" minorities, at the expense of more qualified white individuals and Asians (who now constitute in many cases what has come to be called an "overrepresented" minority). As I have said, this debate has been going on in America for roughly the last forty years. But recently some individuals, whom I will call *squeamish liberals*, have attempted to fashion a new "core argument" for affirmative action. According to squeamish liberals we now need affirmative action to guarantee "diversity" in the workplace and in school. The beauty of the diversity argument as a rationale for affirmative action is that it allows those who deploy it the luxury of simply forgetting about the past. And, in truth, who really wants to bring up that past in polite conversation? Who wants to talk about the horrors of slavery, the torment of the middle passage, the black holocaust, and what Thomas Jefferson (writing in the late eighteenth century but thinking of our America today) referred to as the "ten thousand recollections, by the blacks, of the injuries they have sustained" at the hands of whites?[30] Squeamish liberals certainly do not. And who can blame them? Actually, I think that it would be quite productive if America could get beyond obsessive discussions of its tragic past.

But that is unlikely to happen. Certainly talk of diversity will not bring this end about, regardless of how eager we are to engage in just this talk. Truth be told, diversity simply has no great moral seriousness attached to it. No one fought and died for the sake of diversity. Rosa Parks was not motivated by a desire for diversity in the seating patterns of public buses. Martin Luther King Jr. did not endure police beatings and nights in jail to advance the cause of diversity. Whether one supports affirmative action or opposes racial preferences, the *only* moral issue involved here concerns fundamental fairness. Those who support what they call affirmative action believe that black individuals in particular, but not white individuals, are *entitled* to certain benefits in employment and education precisely because the ancestors of black individuals, but not the ancestors of white individuals, have paid such a high cost in pain and suffering. Thus not guaranteeing affirmative action is akin to cheating blacks out of what is *rightfully* theirs, just as they were in the past cheated out of their individual forty acres and a mule. On the other hand, those who oppose what they call racial preferences believe that the minorities who benefit from such preferences have not paid the requisite cost or met the necessary requirements *today*. Thus those who benefit from racial preferences have cheated by not playing by the same set of rules as everyone else. The point of this debate, again, is simply fundamental fairness. There is no way around this point.

Finally, even the abortion debate may have a lot more to do than most people may like to admit with this larger issue of fundamental fairness. In chapter eight I suggested that the *only* interesting question in the abortion debate is the question of whether fetuses are persons. I still think that is true. But I also think that much of the passion and sense of moral outrage that animates many (though by no means all) opponents of abortion comes simply from the view that women who have abortions are somehow cheating by taking a benefit (sex) without paying a cost (the possibility of a child). Certainly the view that sex is a benefit that comes with a cost could help to reconcile two beliefs which might seem in significant tension one with another: the belief that abortion is wrong and the belief that "artificial" forms of birth control that do not themselves kill a developing embryo—forms of birth control that would therefore make abortion *less* likely—are also wrong. In many respects it is all about costs, benefits, and cheaters.

I began this chapter by asserting that reciprocal altruism, and all of the research surrounding it, represents some of the most fascinating work in the field of evolutionary psychology to date. I hope that the discussion in this chapter has justified that assertion. I also began this chapter by asserting that for nine seasons the popular television show "Little House on the Prairie" provided a weekly lesson in the benefits of reciprocal altruism. Having introduced that show and using it as an extended example throughout this chapter, I cannot conclude my remarks without mentioning the strange fate of Walnut Grove. After nine wildly successful seasons, the writers of "Little House on the Prairie" decided to end the series with a final episode that was truly odd, even by the standards of 1980s television. In that final episode, the residents of Walnut Grove learn to their horror that the very land on which the entire town is situated is actually owned by an outside developer who wants to reclaim the land and everything on it. The residents attempt to fight the developer's actions in court, but are unsuccessful. Nonetheless, those who live in Walnut Grove are a proud people. They decide that while the developer may get the land on which the town is situated, the town *itself* is a different matter. Hence all of the residents, lead by Charles Ingalls himself, choose to dynamite their entire town—to blow up all its residences and buildings, in other words—rather than let it be turned over to an outsider.[31] What if anything this says about reciprocal altruism is anyone's guess.

CONCLUSION

Brave New World Revisited—Again

The public discussion and debate surrounding evolutionary psychology—and, more generally, the return of human nature—has only just begun. But wherever and whenever such discussions take place—in the mass media, in popular-science books, in the academy, at scholarly and governmental conferences—the fundamental issues addressed are strikingly similar to those that have always commanded our attention. Surely, the more things change, the more they remain the same. But the context in which these issues are addressed does change to fit the particular culture or society that is having the discussion. Ours is a world in which—and certainly the United States is a nation in which—the concerns of *liberty* and *equality* set the stage for a good deal of public discourse. In these concluding remarks, I want to venture some final opinions and raise some last questions about the impact that the return of human nature may have on twenty-first-century public discourse. I should hasten to add that the remarks that follow are extremely (perhaps wildly) speculative. Yet if there is to be wild speculation in a book, one supposes that it is best placed in the conclusion.

Let me begin by addressing what may appear to be a curious lacuna in this book. Many readers will doubtless have noticed that nowhere in the preceding pages have I dropped the phrase "brave new world." This was deliberate. In discussions of evolutionary psychology with colleagues and students I have found that inevitably one of my interlocutors will raise the concern that modern science and evolutionary psychology are propelling us headlong toward that dreaded "brave new world." Unfortunately, I have also found that the specter of that "brave new world" often clouds rather than clarifies the important issues under discussion.[1]

Brave New World was, of course, Aldous Huxley's most famous work.

Published in 1932, the novel tells the story of a dystopian future in which humans are genetically engineered to fit within one of five rigidly defined castes, and in which order is maintained by constant social conditioning and chemical "persuasion" that renders the overwhelming majority of individuals quite happy and thus agreeably disposed, but completely unable to realize that they are not free. A large part of Huxley's message was that this "brave new world" will be a dehumanizing place precisely because the struggle—the *agon*—has been eliminated from life. In Huxley's dystopian future, life, liberty, and the pursuit of happiness find their perverted culmination in material consumption without purpose, freedom without responsibility, and the pursuit of mindless pleasures that involve nothing higher than that which is lowest in all of us.

Huxley intended *Brave New World* as a prophecy and a warning. Twenty-six years after that warning, he sought to clarify and intensify his message in a short series of nonfiction articles that were collected under the title *Brave New World Revisited*—a work that functions as something of an "apology" for the earlier novel.[2] Huxley begins *Brave New World Revisited* with the confession that he is, in 1958, even less optimistic about the prospects for humankind than he was when he wrote the novel. Although he calls for more "education for freedom," he seems not altogether sanguine that such education can prove successful at this late hour. He predicts that for the foreseeable future ours will continue to be a world composed of "millions of abnormally normal people, living without fuss in a society to which, if they were fully human beings, they ought not to be adjusted."[3]

Today we have enough historical distance to judge the message presented in *Brave New World* and *Brave New World Revisited*. In general, what strikes me about Huxley's overall vision is its remarkable staying power. Such cannot be said of other dystopian visions that were about in the early and mid–twentieth century. Indeed, Huxley himself contrasts *Brave New World* with that other famous twentieth-century dystopian novel—George Orwell's *Nineteen Eighty-Four*. The contrast is instructive. Orwell's novel is usually read as a dystopia of the political "right," in which totalitarian control is achieved through the terror of the police state. Huxley's novel is usually read as a dystopia of the political "left," in which another type of totalitarian control is achieved through what one might call the distractions of the pleasure state.

I think it fair to say that the impact of Orwell's work has clearly faded with the fall of communism. Fortunately, the more nightmarish aspects of the novel—the telescreens that rob people of any privacy, the secret police with their truncheons, the inhuman torture to which enemies of the state are subjected—now seem fairly remote to us, even though they were quite salient in Orwell's day, when the victory of Soviet totalitarianism seemed an all-too-real possibility.

But if Orwell's novel has lost some of its urgency, Huxley's work may have taken up the slack. It seems so much more immediate—*especially* to intellectuals who feel a special need to "unmask" the lie that many take to be at the root of liberal democracy. I think *Brave New World* and *Brave New World Revisited* get much of their aforementioned staying power from ministering to this need. Indeed, at a *political* level the message of both these works can easily be folded into the contemporary neo-Marxist, postmodern, Foucauldian, and academic feminist critique of liberalism. This critique insists that we are all helpless victims of an all-powerful ideology that offers a sham type of freedom and individuality from which we cannot escape because there is no avenue of dissent that has not already been co-opted by the system.

This may indeed be a "brave new world" of sorts. But it is certainly not a totalitarian world. The element of real control is simply absent. Real totalitarianism is different, because there is a meaningful distinction between persuasion and force. The totalitarian societies about which we need to worry are ones in which terror is palpable, power prances naked in the streets, and no one is even remotely confused about the fact that he or she is assuredly *not* free. There are still too many of these societies on the face of the earth. But they are not Western liberal democracies.

So on a *political-philosophical* level, Huxley's work, although especially popular among intellectuals, may actually have little to tell us that we have not already heard. Intellectuals, after all, have always harbored the sneaking suspicion that the common people may not be quite clever enough to govern themselves. Plato thought this. Huxley may have also. The more things change, the more they remain the same.

But what of the *scientific* prophecies in *Brave New World*? What of the vision of babies in bottles—all clones of one another—waiting to be "hatched" at places like the Central London Hatchery and Conditioning Centre? What of the prenatal conditioning that renders these clones unable to enjoy—or even to desire—working conditions and living environments different from those that they are predestined to fulfill? What of the drugs like soma that, in giving individuals the ability to feel good on demand—without effort or accomplishment—make individuals unable to endure feeling anything less than good? Surely there is something new in this vision that is worth attending to, here at the beginning of the twenty-first century.

Perhaps there is. But I tend to think that even on these points Huxley's vision is flawed. To be sure, it is possible that the military may in the future use cloning technology to create armies of perfect soldiers. And it is possible that multinational corporations may use the same technology to create masses of perfect manual laborers for their assembly lines. It is even possible, I suppose, that some sinister governmental agency might one day dope our water with a somalike substance that makes us all feel perfectly content. Of course, the word would somehow need to go out to members of the agency: "Don't drink the water!"[4]

But I suspect it is more likely that the dangers—or at least the ethical dilemmas—that we will face in the wake of this twenty-first-century science of the self will come upon us as the collective result of free individuals choosing to do what free individuals have always chosen to do: better their lives and the lives of their families. In particular, I suspect that pressure to genetically engineer the "perfect" human will come not from the top down but, rather, from the bottom up. The pressure will come from average individuals who see in technology the possibility that they—or even more likely, their descendants—can be just a little more than average individuals. Ironically, this pressure may need to be restrained by the government.

As neuropharmacological interventions and, later, techniques of genetic engineering become available to effectively treat learning disorders and other cognitive impairments like autism, what individual would not avail himself or herself of this technology? What parent would not want to cure his or her child of an affliction like autism or even of a seemingly mild learning disability? What parent would not want to ensure that he or she did not conceive a child with such afflictions?

I suspect that the answer to these questions is "none," or at least, "not very many." But this leads inevitably to another set of questions that are decidedly more vexing. The difficulty begins when we inquire about the difference between curing a person of an affliction—and thereby returning him or her to "normal" health—and augmenting that "normal" health in some way. Here the issue of what counts as a "learning disability" is particularly on point.

Recently, American schools, universities, and courts have been faced with determining exactly what counts as a legitimate learning disability, amid a growing list of candidates that runs the gamut from the traditional to the truly innovative. Thus, in addition to standard dyslexia—which most people understand as a tendency to mix up letters in words—and Attention Deficit Hyperactivity Disorder (ADHD)—which now has come to mean the inability to sit still and learn in the quiet of the modern classroom—psychologists, teachers, and parents have identified a host of new learning disorders, many with Latin names like dyscalculia (difficulty in understanding mathematical problems), dysgraphia (any of a number of difficulties associated with writing), dyssemia (any of a number of difficulties associated with understanding appropriate responses in a social situation), and dysrationalia (difficulty in being rational).

The problem here is not that there is no bright line between what counts as a *dis*ability and what counts as an *in*ability. The problem is that there is simply no line at all. Over the course of the last decade or so, parents—and well-off parents in particular—have come to this conclusion in vast numbers. The prevailing logic is now that any ability less than a perfect ability is a disability. This logic is clearly reflected in much of the national legislation pertaining to learning disabilities. Thus the Individuals with Disabilities Education Act (IDEA) defines the term *specific learning disability* to mean "a disorder in one or more of the basic psychological processes involved in understanding or in using language, spoken or written, which disorder may manifest itself in imperfect ability to listen, think, speak, read, write, spell, or do mathematical calculations."[5]

This definition raises the tantalizing philosophical question of what would constitute a "perfect" ability to read, or do mathematics, or even, *to think*. It would seem that, on the basis of this definition, we are *all* learning-disabled to some degree, unless there is something more involved in thinking than "one or more of the basic psychological processes involved in understanding."

It does not take a doctorate in education to see where this is heading. If we are all to some degree learning-disabled then grades become not so much measures of what someone knows or what someone can do. Rather, grades become diagnoses of the extent of an individual's disabilities. Very poor grades become evidence of a very great disability. But even mediocre grades become evidence of a less severe disability. The solution is accommodation or correction. The logical result is inescapable: as a society we become like Garrison Keillor's fictional Lake Wobegon, a place "where the women are strong, the men are good-looking, and all the children are above average."

Thus does the logic of contemporary liberal democracy, coupled with the inevitable progress of the science of the self, propel us toward a "brave new world" in which parents, doctors, and well-meaning government bureaucrats come together to help assure the realization of the "perfect" child that every parent just knows is inside his or her seemingly dull, inattentive, combative, or in any way disabled child. Today that assurance comes in the form of accommodations on tests, and also in the form of drug therapies, like Ritalin, that "empower" ADHD children to compete with their less disabled counterparts. In the future, such assurance may come *before* conception by way of genetic engineering. I will repeat that, unless we can draw a line between what counts as a disability and what counts as an inability—a line that I insist we cannot logically draw—the goal of these accommodations, drug therapies, and techniques of genetic engineering will be the production of perfection.

But perhaps there is another kind of line that we *can* draw and that we will choose not to cross. Call this the line between what is natural and what is unnatural. Perhaps we—and here I mean, in particular, parents—will choose to use only "natural" methods to accommodate the disabilities, or even to enhance innate abilities, of our children. There is an analogy in the world of sports. Consider that although we presently have drugs (like anabolic steroids) which can enhance a person's athletic performance, few if any parents of high school football players are giving their sons such steroids. We may be justified in concluding that even if "steroids for the mind" were to become available, few parents would want to see their children take these drugs. The side effects may simply not be worth the risk or the enhancement. Certainly this does seem to be the case with respect to drugs that enhance athletic ability. And, certainly, among those many millions of parents whose ADHD children take daily doses of Ritalin to enhance academic performance, there must be a great many who are more than a little uncomfortable about providing such drugs to their children. Surely all such parents believe that it would be better if those drugs were not necessary in the first place.

It is just that type of thinking that makes genetic engineering decidedly more attractive than crude drug therapies. When the day comes—as come it must—that the technology of genetic engineering makes it possible to identify and correct for at least some of the genes involved in disorders like ADHD, autism, and other learning disabilities and emotional and cognitive impairments, I cannot imagine why would-be parents who have the opportunity to do so would not avail themselves of this technology. I suspect that future parents will jump at the chance to secure the best possible genetic endowments for their children. But what does "best" mean in this "brave new world"? Why disadvantage your child by passing on to him or her only a mediocre intelligence, when you can give him or her a "head start" in this most competitive world by bumping up his or her innate intelligence a standard deviation or two? Why conceive a follower, when you can conceive a "born leader"? I suppose that I need not point out that future parents will be addressing these questions within a context in which at least some *other* future parents have decided to "perfect" their own children.

At this point a cry will rise up in the land. This cry may amount to a demand for the immediate halt to any future research into genetic engineering and perhaps for laws against the use of such genetic engineering technologies as are currently available. But the benefits of genetic engineering in curing real and significant disorders will simply be so great that, I suspect, this particular cry will *not* be heard. Rather, the dominant cry will be in the form of a demand that *every* prospective parent be guaranteed the opportunity to share in the benefits of genetic engineering technology regardless of his or her economic or class standing. It will be insisted that this must not be a technology only for the rich, or the educated, or members of the "majority" race. As happens to all good things in a liberal democracy, the benefits of genetic engineering technology will inevitably become an *entitlement*. But as with all entitlements, there will be strings attached, in the form of governmental regulation and administration of these benefits. Thus I predict that in this century the United States and other Western liberal democracies will adopt laws that regulate procreation in very specific ways. These laws will regulate not only the number of children individuals may produce, but also the types of genetic enhancements that can be made available to these children, as well as the types of genetic maladies that must be eliminated from the genome of these children. All of this, I insist, will come about through normal democratic means. The result will be what I call a "*democratic* eugenics."

I think that such a democratic eugenics is inevitable, although not for the reason that is typically advanced. Typically, it is argued that we must control births in both developing and developed countries, lest we face the nightmare that awaits us when the population bomb explodes. Bluntly put, the argument is that since resources are ultimately finite we must stop procreating at our current rate or we will all die early deaths. I do not wish to debate the merits of this argument, except to say that for every prophet of doom who predicts that the world will end in eco-catastrophe, there is a prophet of progress who assures us that technology and human ingenuity will save the day.[6] But I will insist that whatever the truth of this argument, the language of scarcity and limits will never bring about a democratic eugenics. Such language may, however, bring about a forced and totalitarian eugenics—a world where fundamental freedoms really are denied.

To bring about a democratic eugenics, however, one needs to speak the language of democracy: liberty and equality. With respect to equality, I think that the case for a democratic eugenics is strong today, and will become irresistible in the future. As significant advances in genetic engineering technology start to be made, those who are not wealthy enough to avail themselves of this technology will naturally insist that it is an entitlement which must be made available to them regardless of their ability to pay. Call this "genetic security." In short order, this significant constituency may well be joined by the broad middle class majority, which after reexamining the connection between liberty and responsibility, may come to the belief that although individuals may be entitled to have a limited number of children (and perhaps to have the government ensure the genetic and financial security of some of these children), this entitlement does not extend to individuals who willfully attempt to have more children than they can support, or than the government has agreed to support.

We are slowly but surely moving down this road. Already there is court precedent for restricting procreation because of an individual's consistent and willful disregard for his legal obligation to provide any support whatsoever to children he has already helped to produce. In the summer of 2001, the Wisconsin Supreme Court upheld a lower court's order that, as a condition of his parole after being found guilty of a felony violation for failure to provide child support, David Oakley be restricted from conceiving another child with any woman. This restriction applies while Oakley is on parole, or until such time as he can make good on his outstanding obligations to provide child support to the nine children he has already fathered with four women. In this case, the court's goal seems to have been to deter sexual irresponsibility, not to deter conception by poor individuals.[7] Still, the practical effect of the decision clearly "discriminates" against poor individuals. To prevent this, in the future the government may need to guarantee what one might call a *minimum right to procreate*. Once this is done, some significant restrictions on procreation may logically follow.

Although the United States Supreme Court itself has found that procreation is a "basic liberty," the type of democratic eugenics I have sketched is not inconsistent with this liberty.[8] As any first-year law student can tell you, no liberty is absolute. Your right to do with your own body as you please, so long as you do not adversely affect another body, does not seem logically to extend to your *production* of another body. Whether that production comes about through the usual means of human conception, or by any of the new methods (cloning or whatever else we will shortly develop) by which something that is recognizably human can be brought into existence, there is necessarily another nonadult body—hence one that cannot give consent—to be considered. Surely society has some interest in that nonadult body. That interest is immeasurably strengthened when the choice facing adults of child bearing age becomes not just *whether* to conceive a child, but *what type* of child to conceive. A democratic eugenics would not reach into every facet of this choice. But it might be able to put certain restrictions on this choice, without becoming unduly burdensome.

I realize that my remarks about a democratic eugenics, though speculative, may appear highly controversial. Doubtless there are those who will insist that no eugenics program of any sort could ever be democratic, and that if such a program were attempted in America this nation would no longer be the land of the free or the home of the brave. I am sympathetic to this point. Today it may seem difficult to envision a free America that also embraces what I have called a democratic eugenics. Thus to give my vision some degree of plausibility, I want to engage the reader in a thought experiment involving time travel.

Suppose that you were able to travel back in time to (say) New York City in the year 1840. Suppose that on a street in lower Manhattan you met an average male citizen (say, a forty-something-year-old shopkeeper), whom you proceeded to engage in conversation. Suppose that you told this shopkeeper that you were from America's future, and suppose that you then proceeded to describe that future as it would exist in the year 2002. You would be eminently justified in telling your interlocutor that twenty-first-century America will be a dramatically more free, more equal, more diverse, and more secure place than it is in 1840. You could tell him that in the next 162 years Americans will have abolished slavery—albeit as the result of a civil war—but also that Americans will have *democratically* decided to give every adult citizen the right to vote regardless of the citizen's color or sex, to abolish child labor, to secure old-age pensions to all workers, to provide food, medical care, and other necessities to the poor, and in general to make citizens (and even noncitizens) materially better off than anyone in the early nineteenth century could imagine. An American hearing this in 1840 would surely be happy to envision his or her descendants living in twenty-first-century America.

But now suppose that in the course of this conversation you also happened to mention to this average New Yorker that in the twenty-first century one of the largest governmental agencies in America would be something called the Internal Revenue Service. Suppose you noted that this agency would keep detailed and extensive records on every American, records that included information on where the individual lived, where he or she worked, whether he or she was married, how many dependants he or she had, how much money he or she made, what specific assets he or she owned, how much money if any he or she derived from these assets, and even how much he or she gave to charity. Doubtless your interlocutor would be at least mildly surprised at the extent of information on Americans that was possessed by this huge governmental agency. He might ask what nefarious and all-pervasive surveillance methods this agency used to acquire this information. You would then need to explain that all this information is *voluntarily* provided, either by the individual himself or herself or by the individual's employer. You might add that this information is required by law. But you would then need to say that, on the whole, the overwhelming majority of Americans in the twenty-first century are more than willing to provide the information in question without the least protest.

Your interlocutor might then ask you the purpose for which this information is gathered. You would need to inform him that it is used for the purpose of determining how much annual federal *income* tax each American must pay. Now, *in general*, taxes and other "duties" and "fees" are certainly something that an American in 1840 could well understand. Even in 1840, governments at all levels needed money to build roads and schools, to provide for the common defense, and generally to get on with the business of the government. Of course, your interlocutor would know that there are numerous methods other than an *income* tax by which the federal government could raise money to carry on its operations: methods like a direct sales tax, import duties, and other fees, many of which were in use in the mid–nineteenth century. Still, your interlocutor might reasonably assume that the Internal Revenue Service and the federal income tax were principally designed to collect taxes for the purpose of raising money for the type of government that would be needed in a large and complex industrial nation. But if your interlocutor did assume this, you would be honor bound to disabuse him of this belief. You would need to point out that the purpose of the federal income tax is principally to engineer a particular type of society by reaching directly into the lives of citizens and controlling those lives in such a way that the citizens will come to do what the government has democratically decided that they should do. The result, you might explain, is a federal income tax code that is thousands of pages long and that is frankly incomprehensible to the vast majority of Americans—yet a code that affects the life decisions of each and every American on an almost daily basis. You might note, for example, the many and varied ways that the twenty-first-century American federal income tax code uses monetary rewards and punishments—bribes and fines, essentially—to induce Americans to behave in certain ways. Your list could include the following twenty-first-century tax realities: If you choose to spend a thousand dollars a month for a mortgage on a house in the middle of Peoria you are favored by the tax code and given what amounts to a generous subsidy, but if you choose to spend the exact same amount of money for rent on an apartment in the middle of Manhattan you are not favored by the code. Also, if you save money for your retirement so that you will not be wholly dependent on the government in your old age, you may be favored by the tax code, but if you save *too* much money you will not find additional favor and you may even be penalized for putting that additional money toward savings rather than consumption. Such, you might say, is life in twenty-first-century democratic America.

At this point your interlocutor would surely raise the issue of constitutionality. He would doubtless point out that the kind of tax code you are describing is directly prohibited by Article I, section 9, of the United States Constitution, which clearly forbids the kind of income tax mischief and attendant social control that you are describing. You would need to concede to your interlocutor that he is, in fact, correct on this point of constitutional law. But you would then need to inform him that, in 1913, the following one-sentence amendment was added to the United States Constitution: "The Congress shall have power to lay and collect taxes on incomes, from whatever source derived, without apportionment among the several States, and without regard to any census or enumeration."[9] Dumbfounded by the almost infinite scope of this language, your interlocutor might inquire as to where the *intellectuals* were during this whole period, for he might confidently assume that the intellectuals would explain to the general public what a tax code like the one you have described would do to their most basic liberties. You would need to explain that in general it is the intellectuals, more than any other single group, who support the establishment and maintenance of a progressive tax code that rewards certain types of activity more than other types and that is designed to affect the decisions and lives of everyday Americans in specific and complex ways. You might also point out that almost all of these intellectuals support such a code not because they think it undermines democracy or individual freedom but, rather, because they sincerely believe that it enhances these values.

I think the intellectuals are correct in this belief, although there are surely those who would disagree. But one thing is certain. For at least the last fifty years, since the federal income tax has become a significant part of most Americans' lives, there has been abroad in the land a robust debate about the tax code itself, and in particular about how modifications in the code—"tax breaks" as they are now called, even by those who would not recognize a page of tax code if they saw it—should be used to help us become the type of people that we democratically decide we should be. I am suggesting that in the not too distant future we will be having this same type of discussion about matters concerning procreation and genetic engineering. As with our discussion regarding tax policy, this future discussion will also implicate vital issues that go to the core of what type of human beings we democratically decide we should be. I think this discussion can be fruitfully carried on without reference to, or worry about, the specter of a "brave new world."

This discussion, however, may still be several decades in our future. Of more immediate concern is the discussion we are now having regarding the very essence of who we are—a discussion that has been occasioned by the return of human nature from a century of exile. In the previous pages I hope to have shown how evolutionary psychology is contributing to that discussion by changing the way we talk about ourselves and our world in the twenty-first century.

Since this has been a book about evolutionary psychology (the theory) and public discourse (the practice), I want to end on an ironic note about the relationship between the theory and the practice. You will recall that in the introduction I defined public discourse as public debate and discussion on significant policy questions. Even when we emphasize the activity of discussion over debate, there is always a *persuasive* aspect to public discourse. Now, suppose we follow Aristotle and say that *rhetoric* concerns itself with the discovery of the available means of persuasion in the given case.[10] Suppose we then assume that the available means of persuasion in any case will be constrained in part by characteristics of human nature. Any good theory that can shed more light on human nature should then be able to shed more light on the possibilities and limits of persuasion. Evolutionary psychology holds the potential of being able to do just this. In fact, by drawing on the insights of evolutionary psychology it may be possible to develop a wholly new rhetoric of public argument—one that is more firmly grounded than previous rhetorics in a tenable view of human nature. Call this an adapted rhetoric of public argument. Such a rhetoric will help us to understand ourselves more fully, by helping us to understand why certain issues seem so important to us, and why certain arguments seem so *naturally* persuasive to us. I suspect that such a rhetoric will help us see that those who are most successful at public discourse in a democracy are, in a manner of speaking, orators of the Pleistocene. They are the ones who best understand our evolved mind. That understanding is important for all of us to possess, as we begin our public discourse on the most important issue our species has ever addressed—the issue of what type of species we want to become. At the very least, such an understanding may help to sharpen the bounds of the politically possible, and in so doing point the way toward a more productive and progressive democratic politics than the one that intellectuals have been able to manage since the 1960s. An adapted rhetoric of public argument may ultimately take us as far as we can go in perfecting our species through language. But that is the subject of another book.

AFTERWORD

Writing on *The Blank Slate*

Just after this book went to press, Steven Pinker published *The Blank Slate: The Modern Denial of Human Nature*.[1] I strongly suspect that one of Pinker's principal motivations for writing his new book was to correct what he sees as often robust misunderstandings (both scientific and political) regarding evolutionary psychology. Pinker finds numerous misunderstandings in need of much correction. *The Blank Slate* is thus a large book, not just in the sheer number of its pages but also in the scope of the material covered, and in the significance of the claims made. The most important of these claims is that our modern denial of human nature is both *scientifically untenable* and *politically detrimental* to the progress of liberal democracy.

To demonstrate that this denial is scientifically untenable, Pinker reviews some of the central concepts in evolutionary psychology and discusses some interesting findings regarding the workings of the mind. Readers of my book and of Pinker's earlier books will find some of this material familiar, but Pinker also adds new details. In particular, he wants to make certain that no one could accuse *him* of not keeping up-to-date regarding matters of the mind. Thus he spends considerable time discussing, and ultimately attempting to debunk, some of the more grandiose claims made by adherents to one of the hottest new ideas in neuroscience: neural plasticity. Assessing this whole idea and its relationship to the nature-versus-nurture debate, Pinker writes:

> Discoveries of how the brain changes with experience do not show that learning is more powerful than we thought, that the brain can be dramatically reshaped by its input, or that the genes do not shape the brain. Indeed, demonstrations of the plasticity of the brain are less radical than they first appear: the supposedly plastic regions of cortex are doing pretty much the same thing they would have been doing if they had never been altered. And

the most recent discoveries on brain development have refuted the idea that the brain is largely plastic.[2]

This claim chimes perfectly with everything Pinker has said in his previous books and with his larger argument, pressed hard in his new book, that, far from coming into the world as a blank slate, every normal human mind is equipped with a collection of innate cognitive faculties and core intuitions.

Notice that this is a factual claim about the physical world. Could Pinker and other evolutionary psychologists be wrong about this claim? Indeed they could. And although it may seem paradoxical, it is absolutely critical to the argumentative success of evolutionary psychologists like Pinker that they continuously reaffirm their claim that they could, in fact, be wrong about their factual claims about the physical world. But how would we *know* if they were wrong? Pinker's answer is that we can only know this by way of *real* science. This is what Pinker means when he rather boldly states, "Human nature is a scientific topic."[3]

There's the rub! Pinker's real argument with fuzzy-minded intellectuals who deny the existence of a universal and knowable human nature, or who think that the mind is infinitely plastic, is that they do so *not* on good scientific grounds but rather on the basis of what are essentially political motives. In twenty-first-century American culture, the contrast between "scientific grounds" and "political motives" is as invidious as it gets, and Pinker uses this contrast to great effect throughout *The Blank Slate*. The contrast is between motivation that is disinterested and thus reliable and motivation that is interested and thus suspect. The latter motivation is suspect precisely because motivation that is interested is always interested in *power*. Science, on the other hand, remains true to itself—remains real science—only to the extent its ultimate interest is truth and its only authority is reason.

But reason and power, truth and political interest need not be enemies. Indeed, the whole of the Enlightenment project was an attempt to unite science and politics for the betterment of humankind. Thus in the preface to *The Blank Slate* Pinker, sounding like a twenty-first-century Diderot or Condorcet, writes, "The new sciences of human nature can help lead the way to a realistic, biologically informed humanism."[4] In writing this, Pinker is carrying the torch of his Enlightenment forebears, even if he does not cultivate that characterization as actively as do some of his older colleagues—notably Edward O. Wilson. The characterization is accurate, however, and it draws our attention to an important point. Although it does not announce the fact, *The Blank Slate* is really as much about the place of science in society as it is about the relationship between evolutionary psychology and politics.

Even so, the explicit focus of *The Blank Slate* is on how scientific advances in the disciplines of evolutionary psychology and cognitive science are being obscured or even undermined by those fuzzy-thinking intellectuals who refuse to see the facts of human nature but instead embrace politically motivated doctrines that amount to no more than pseudoscientific myths. Thematically, the book is concerned with three such doctrines, each of which has attained "a sacred status in modern intellectual life," but each of which must, according to Pinker, be debunked.[5] These three doctrines are: the Blank Slate, the Noble Savage, and the Ghost in the Machine. While not wishing to impose upon Pinker's book a structure that its twenty chapters do not present, I think the most effective way of describing what he is trying to do is to discuss each of these three doctrines in order, weaving into these discussions parts of the book that are relevant to the doctrine being discussed.

As one would expect from his title, the doctrine of the Blank Slate receives by far the most attention in Pinker's book. This doctrine is today most closely associated with the seventeenth-century philosopher John Locke, although Locke himself used a different metaphor to connote the essential idea that the mind comes into the world without any ideas. Pinker openly and sometimes derisively wonders why so many otherwise intelligent individuals have come to believe a doctrine that is so evidently scientifically false. He provides four explanations, devoting a chapter to each.

One explanation involves the "fear of imperfectability." The idea is that if the mind does come with innate desires, attitudes, intuitions, and so forth, then it may be resistant to "educational" attempts to "improve" it. Pinker nicely turns this entire discussion around by noting that, if history is any guide, we ought not to fear imperfectability. Rather, we ought to fear those who say that they *can* perfect human beings. Such individuals may begin as utopian dreamers, but if given power, they inevitably end up as totalitarian dictators. As Pinker understatedly notes, "the *denial* of human nature can be *more* dangerous than people think."[6]

Two other explanations as to why intellectuals embrace the doctrine of the Blank Slate involve the "fear of determinism" and the "fear of nihilism." These fears are obviously related. Their force depends upon the assumption that if the mind is not a blank slate, then our biology must, to some extent at least, determine our actions and our will. So-called biological determinism is obviously a threat to free will, and where there is no free will there can be no meaning to life; hence the void and nihilism.

Pinker provides answers to both of these fears. I will come to those answers when I discuss the last doctrine that Pinker wants to debunk. For now, I want to concentrate on what is by far the most important explanation Pinker advances as to why intellectuals embrace the doctrine of the Blank Slate: the "fear of inequality." The idea here is that no belief in political and social equality can be sustained without a corresponding belief that human minds come into the world as blank slates. Since liberal democracies *must* believe in political and social equality, they must also believe in the doctrine of the Blank Slate.

If Pinker is to be successful at debunking this doctrine, he must alleviate this fear *completely*, because inequality is the one evil that liberal-democratic cultures simply cannot tolerate. His direct and forceful attempt at eliminating this fear centers on the only two types of political and social inequality that really matter in twenty-first-century American society: the types based on sex or race.

Regarding evolutionary psychology's relationship to our liberal-democratic concern for equality between the sexes, Pinker stresses that males and females share the same basic architecture with respect to the design of most parts of their brains. This fact helps to ratchet down the level of anxiety that surrounds discussions of sexual equality, because it means that men and women are, by nature, equally well qualified for most jobs—certainly most jobs in post-industrial, information-based societies.

The problem, however, is that the parts of the brain that do differ between men and women—parts involving mate selection and child rearing, at least—play a very significant role in all people's lives. Thus Pinker follows Kingsley Browne and others in suggesting (as I noted in chapter nine) that the reason women as a group earn some fraction of a dollar for every dollar earned by men is not just because women face a "glass ceiling," but also because some women voluntarily take lower-paying jobs that allow them to spend more time with their families. As a society, we could correct for this imbalance by utilizing hiring quotas based on sex, or by trying to indoctrinate young women to believe that they have been indoctrinated to believe that they prefer to be with their children rather than attend a business meeting. The first solution might work; the second has been tried and has utterly failed. But neither solution, according to Pinker, is necessary or just in a liberal democracy. Therefore, in a decisive move, Pinker asserts that justice in a liberal-democratic society demands *only* that we guarantee "equality of opportunity." Indeed, as Pinker repeatedly stresses in a number of contexts, it is unjust to use coercive state pressure to enforce numerical equality between the sexes in any domain of life if the inequality is not the result of prejudice.

This nicely takes care of the "fear of inequality" argument as it relates to sex. Pinker is saying that, rather than fearing what is natural, we ought to fear misguided and often coercive governmental attempts to make what is unnatural seem natural. But notice that this same approach could work for the "fear of inequality" argument as it relates to race. One could argue that there are significant differences between blacks and whites in the structure of their brains, and that these differences account for some of the differences I discussed in chapter two—particularly the differences in average IQ test scores and in average incomes between blacks and whites. One could then simply insist that—as with sex, so with race—justice demands only that we guarantee equality of opportunity to all.

This approach would, however, be a disaster for evolutionary psychologists. As a society we *are* willing to accept that there are significant emotional and psychological differences between the sexes that are not socially constructed. But we are absolutely unwilling to accept that there are significant, innate, *qualitative* differences between blacks and whites—especially in the realm of the mental. Fortunately for evolutionary psychologists, we do not need to accept this belief, since it is scientifically false. Thus, in a critical passage that refers specifically to race, but *not* sex, Pinker writes: "People are *qualitatively* the same but may differ *quantitatively*. The quantitative differences are small in biological terms, and they are found to a far greater extent *among* the individual members of an ethnic group or race than *between* ethnic groups or races. These are reassuring findings. Any racist ideology that holds that the members of an ethnic group are all alike, or that one ethnic group differs *fundamentally* from another, is based on false assumptions about our biology."[7]

The statistical point Pinker makes regarding the greater variance of *certain* traits among individuals in certain groups than between averages of those groups harkens back to my discussion in chapter three regarding Richard Lewontin's blood factor analysis of genetic differences between the races. This analysis is complex and often misunderstood. But I think Pinker would agree that it does not compel the conclusion that, although the races may share the same basic brain structures, the *quantitative* differences in the operation of various brains cannot be statistically significant, heritable, and important for overall success in any society. To put it bluntly, nothing in anything I have read by Pinker absolutely refutes the scientific hypothesis that *some* of the black-white difference in IQ is attributable to biological differences between the races. To be sure, Pinker himself does reject this hypothesis. He follows Thomas Sowell and others in expressing the "view" that *all* of this difference can be accounted for based on environmental factors alone. But the fact that Pinker does *not* present this view as a scientific fact may actually make *The Blank Slate* more persuasive (because more honest) than it would be if he were to insist on a claim that science has yet to establish. Still, Pinker's less than thorough approach to the vexing issue of race and intelligence in his latest book continues the policy of benign neglect that I have already suggested seems to characterize the work of many evolutionary psychologists on this matter. Sometimes his silence on this issue is deafening. Consider, for example, the last section of *The Blank Slate*, entitled "Hot Buttons." Here Pinker devotes entire chapters to the relationship between evolutionary psychology and some of the most controversial political issues of our time, including sexual equality, the causes of violence, and even the naturalness of capitalist economic systems. But conspicuous by its absence is a chapter on what is certainly the hottest of all the hot button issues: the question of race and intelligence.

Pinker is perhaps wise not to engage this issue directly. Instead, throughout *The Blank Slate* he repeatedly maintains that discrimination would be wrong even if we did find statistically significant biological differences between races. He advances the simple argument that discrimination is wrong just because it causes one to judge "an *individual* according to the average traits of certain *groups* to which the individual belongs."[8]

This argument embodies the essence of the mainstream American position on justice and fairness. But there is a problem that Pinker immediately recognizes: Often it *is* rational to judge individuals according to traits that can be associated with groups to which they appear to belong. For example, if you see a man in a business suit and a woman holding an infant, and if you want to know the price of a box of Pampers but can only ask one of these individuals, it would be rational to ask the woman. This rational solution might fail you. It could be the case that you have stumbled on a situation in which the woman knows nothing about the price of diapers because she is a business executive on vacation who leaves her infant child with her stay-at-home husband, while the man in the business suit could answer your question immediately because he is a sales representative for Procter and Gamble. But in twenty-first century American society this does not make your initial decision an example of irrational discrimination. You chose between two individuals based on the best evidence available to you. If you had had better information, you could have made a better choice. The same could be said of employers or educational institutions, when it comes to choosing among individuals of different races or sexes. Thus, in what is my candidate for the most provocative passage in *The Blank Slate*, Pinker writes: "The best cure for discrimination, then, is more accurate and more extensive testing of mental abilities, because it would provide so much predictive information about an individual that no one would be tempted to factor in race or gender. (This, however, is an idea with no political future.)"[9]

Pinker is no doubt correct that more extensive mental testing is unlikely in today's political climate. But I cannot help thinking that the above passage misses the larger point. Even if we grant that more extensive mental testing would mean that we would not need to "factor in" race or sex when considering an individual's qualifications for a job or an educational opportunity, this still does not mean that we would not *notice* that individual's race or sex.

Now, noticing an individual's sex is, I think, unproblematic because as a society we are willing to believe that men and women are different, and that these differences matter, but that these differences need not imply that one sex is superior to the other. In truth, human nature has assigned different abilities and roles to males and females—at least with respect to *some* significant aspects of life. As a liberal-democratic society we can get along quite well by simply guaranteeing equality of opportunity to all, without regard for sex, while also regarding each sex as equally important, because each is directed toward a *common* goal—the continuance of the species. No one's dignity is diminished in this view.

But the same is not the case for race—even if we assume that differences in race are merely quantitative. Even if we do not *need* to factor race into employment or educational decisions, simply to *say* that a certain individual's superior performance on a job or a test is exceptional "for his race" is plainly demoralizing to *all* members of that race. It also eats away at the necessary foundation of fraternal feeling in any society. It invites talk of natural superiority and inferiority—and, in its ugliest manifestation, talk of master and slave—in a way that sex role discussions need not. In short, nature simply cannot be assumed to have assigned different abilities to the races, though it can be assumed to have assigned different abilities, and even different roles, to the sexes. Thus unless and until modern science finally slams the door on the claim that there is any qualitative *or* quantitative difference at all between the races in terms of intelligence, the "fear of inequality" will continue to be a powerful motive for embracing the doctrine of the Blank Slate. Try as he might, Pinker is not successful at allaying this fear.

So much for the doctrine of the Blank Slate. The next doctrine Pinker attempts to debunk—the Noble Savage—has entered our current cultural vernacular by way of the cultural left's interpretation of Rousseau's philosophy. Rousseau was a uniquely subtle philosopher. The cultural left reads him as saying that man in the state of nature was noble and peaceful, without greed or envy; hence we should all strive to return to the state of nature. It would be wrong to call this a bastardization of Rousseau's thought, for it is simply not Rousseau's thought at all. Rousseau was emphatic that we cannot go home again. There is no *simple* return to the state of nature. Donning expensive designer "peasant attire," munching granola, and protesting the sale of genetically engineered food does not bring one closer to the state of nature. These are sham attempts at "authenticity" that reveal the kind of amour-propre that Rousseau so detested. If you doubt this, just consider the many health food fanatics who live in fear that they might ingest a muffin made from genetically modified corn, but who take daily doses of oral contraceptives. The message here is clear: We must get back to nature—except when it would be inconvenient to do so.

This "back to nature" message—however contradictory it is in practice—has nonetheless become a kind of rallying cry for the cultural left. But it can have seriously detrimental political consequences. Thus Pinker laments the fact that "school-children are currently fed the disinformation that Native Americans and other peoples in pre-state societies were inherently peaceable, leaving them uncomprehending, indeed contemptuous, of one of our species' greatest inventions, democratic government and the rule of law."[10]

The simple truth, as Pinker points out, is that far from being a collection of "noble savages" pre-state societies (like the Yanomamö, a Brazilian tribe that the anthropologist Napoleon Chagnon made the object of thirty years of study) tend overwhelmingly to be more warlike, xenophobic, paternalistic (if not misogynistic), and just plain dangerous than the worst liberal-democratic society on earth. The point is not that the Yanomamö tribes are composed of biologically aggressive or violent people. They are not. The point is that, because of cultural inventions like representative government, private property, contracts, and the rule of law, liberal-democratic societies (of which all people are biologically capable) are the safest, freest, most egalitarian societies on earth. For just that reason, they are also the best places on earth for the weak, the poor, women, and minorities. The fact that school children (and college students as well) are routinely taught the reverse is, as Pinker makes clear, a perverse effect of the doctrine of the Noble Savage.

This brings us, finally, to the doctrine of the Ghost in the Machine. While the first two doctrines are disproportionately embraced by those on the cultural left, the doctrine of the Ghost in the Machine is probably disproportionately embraced by the cultural right and religious individuals in general. As Pinker notes, we owe the phrase "the Ghost in the Machine," to the twentieth-century philosopher Gilbert Ryle, who used the metaphor as a way of deriding the Cartesian belief that there simply must be some immaterial substance in the human body or mind that is absolutely unaffected by any of the laws of nature. The belief that there must be such a substance—call it a spirit or a soul—is generally thought to be a necessary condition for the existence of free will. Thus to believe that such a substance does not exist leads, as Pinker says, to the fear of determinism and the fear of nihilism. Debunking the doctrine of the Ghost in the Machine, without raising the two specific fears that its embrace is meant to allay, involves Pinker in some dexterous refutation that takes place both on a practical and metaphysical level.

On a practical level, it is often argued that without some notion of free will the notion of personal responsibility becomes untenable, and hence society unravels. This is particularly true in considerations of issues involving crime and punishment. If people really are just machines, all the way down, then it does seem at least counterproductive, if not cruel, to punish them for actions that they could not control. In this view, rapists and cancer patients are essentially similar. Both are suffering from diseases and need treatment (and perhaps even sympathy) rather than punishment.

This view is widely embraced by intellectuals on the cultural left. But after inveighing against the positions of these same intellectuals on issues concerning equality of opportunity and the peacefulness of life in pre-state societies—and indeed after calling into question the intellectual rigor of such positions—

it would be uncomfortably awkward for Pinker to embrace the position that punishment is unjust because most criminals are not responsible for their actions. To avoid this awkwardness, Pinker must find a way of shoring up the concepts of responsibility and punishment without tying them to the idea that there must be a ghost in the machine.

To do so, Pinker first advances a pragmatic argument. He insists that in order to continue using concepts like responsibility in the context of punishment for crimes, "we only have to think clearly about what we want the notion of responsibility to achieve. Whatever may be its inherent abstract worth, responsibility has an eminently practical function: deterring harmful behavior."[11] Pinker's point here seems to be that, like the concepts of representative government, private property, and the rule of law, the concept of personal responsibility is in some sense a "fiction"—but a very useful one for liberal democracies. By "pretending" that *most* criminals are responsible for their crimes, we deter future crimes more effectively than we otherwise could, and we appease a general sense of "justice" that may well be hardwired into our brains.

On a practical level, that analysis might suffice. But in a quiet hour, when the streets are safe and the time for contemplation is at hand, the mind turns back to questions of the "inherent abstract worth" of responsibility, especially as it relates to justice. Even if the safety of society can best be maintained by assuming the existence of personal responsibility and then utilizing the threat of punishment, can that threat be grounded on anything "higher" than a simple call for "law and order"? If not, then justice becomes simply the exercise of power in the service of gaining more power. But this sounds a lot like "survival of the fittest." Assuredly, there is a kind of nihilism connected to this view. I take it from his book that this is not the view Pinker wishes to endorse.

On a metaphysical level, then, Pinker needs to tie moral responsibility to justice in a way that does not involve positing a ghost in the machine but somehow holds the promise that life is about more than simply power. He needs ultimately to advance his own utopian vision of the universe, if his book is to be persuasive. Pinker does so in spectacular fashion. He begins by following (he says) Plato and other "moral realists" in asserting that there are some concepts in the universe that exist independent of human thought. The concept of number is an example. Presumably, any intelligent alien civilization we would ever find would know, just as we do, that two plus two equals four. That truth is not unique to the human species. It exists in the universe, waiting to be discovered by any being that possesses intelligence.

Pinker then argues that, as it is with the arithmetic, so it may be with justice. There are certain moral truths that exist independent of humans and that command our rational adherence because, *given the structure of the universe*, it would be self-defeating to act contrary to these truths. As Pinker explains:

> The world presents us with non-zero-sum games in which it is better for both parties to act unselfishly than for both to act selfishly (better not to shove and not to be shoved than to shove and be shoved). Given the goal of being better off, certain conditions follow necessarily. No creature equipped with circuitry to understand that it is immoral for you to hurt me could discover anything but that it is immoral for me to hurt you. As with numbers and the number sense, we would expect moral systems to evolve toward similar conclusions in different cultures or even different planets. . . . Our moral sense may have evolved to mesh with an intrinsic logic of ethics rather than concocting it in our heads out of nothing.[12]

The argument Pinker makes here echoes the entire discussion of reciprocal altruism that I presented in chapter ten. There I noted that a non-zero-sum environment is one in which any individual's gains could be greater than the gains he could achieve alone, and no individual need necessarily suffer a loss because of the gains of anyone else. In this environment, mutual cooperation is the best strategy for everyone.

Pinker suggests that the world contains many such environments. Hence it is appropriate and necessary to adopt the particularly appealing moral stance he advocates—essentially, the Golden Rule—in these environments. But is the universe *itself* such an environment? If not, then the particular morality in question is no longer universal but is instead only local and parochial. Even worse, this now parochial morality might not fit our particular *world*, if our entire world is not essentially a non-zero-sum environment.

Could our world possibly be such an environment? It is difficult to think so, given the problem of finitude. Material resources are scarce, after all. This is usually taken to mean that there are limits to growth—including the growth of the human population—and that when these limits are exceeded the inevitable result is war, famine, pestilence, and death. In the late seventeenth and early eighteenth centuries, Thomas Malthus predicted just such a future, when food supplies will be depleted by an exponentially increasing human population. Of course, mutual cooperation is still *possible* in a Malthusian world. But this cooperation takes place against the backdrop of a now tragically configured moral universe in which the price of life for some must be the death of others. Thus cooperation, though still laudatory, may not be *rational* for a given individual in a zero-sum environment.

A great deal is therefore riding on the type of universe we inhabit. For Pinker's claim about the objectively true moral status of the Golden Rule to be valid, ours must be a non-zero-sum universe. To make the case that it is, Pinker turns to the work of the contemporary economist Paul Romer. As Pinker explains, Romer begins his argument that we live in a non-zero-sum world "by pointing out that human material existence is limited by *ideas*, not by stuff. People don't need coal or copper wire or paper per se; they need ways to heat their homes, communicate with other people, and store information. Those needs don't have to be satisfied by increasing the availability of physical resources. They can be satisfied by using new ideas—recipes, designs, or techniques—to rearrange existing resources to yield more of what we want."[13]

Ideas are the key to the equation. Pinker and Romer clinch the argument by noting that, although it is *usually* the case that if I cooperate and share something I have with you, I necessarily have less of whatever I have shared, this is *not* the case if I share an idea. Thus ideas conquer material limits, but there are no limits to ideas. This is how we can believe that reciprocal altruism as a moral command is woven into the very structure of our universe.

The reader who has been indulgently following this "metaphysical" discussion may well wonder exactly what all of this has to do with the task of debunking the myth of the Ghost in the Machine. The properly metaphysical response is to note that, because reason and science tell us that we live in a non-zero-sum universe in which reciprocal altruism is the best policy for everyone, reason must direct the will toward the common good. Thus there is no need to posit a ghost in the machine—a ghost that chooses between good and its opposite—because in the end there is no place for choice. This is a type of determinism, to be sure. But it is not to be feared, since the will is not determined by our biology *as such* but rather by the essential structure of a universe in which mutual cooperation brings about limitless good. Mutual cooperation is thus good *and* morally imperative. This is how one ties the concept of moral responsibility to justice in a way that does not involve positing a ghost in the machine, but somehow holds the promise that life is about more than simply power. I would also note that this is essentially Plato's view of the relationship between the will and justice—embodied in Plato's famous comment that "the just man will never wish to do injustice," and captured more generally in the Platonic notion that knowledge *is* virtue.[14]

In the end, the vision of a realistic, biologically informed humanism that emerges after the doctrines of the Blank Slate, the Noble Savage, and the Ghost in the Machine have all been debunked looks a lot like contemporary liberal democracy, in which Judeo-Christian ethics have been fully secularized and science and technology lead the way toward expanding vistas of material and social progress. Such is the vision that Pinker presents in his latest book. It is not an unpersuasive vision.

NOTES

Introduction: "This Changes Everything"

1. Richard Rorty, *Contingency, Irony, and Solidarity* (Cambridge: Cambridge University Press, 1989), xiii.

2. Steven Pinker, *How the Mind Works* (New York: W. W. Norton, 1997), 21.

3. Jerome H. Barkow, Leda Cosmides, and John Tooby, "Introduction: Evolutionary Psychology and Conceptual Integration," in Jerome H. Barkow, Leda Cosmides, and John Tooby, eds., *The Adapted Mind: Evolutionary Psychology and the Generation of Culture* (Oxford: Oxford University Press, 1992), 3.

4. Evolutionary psychologists, including the editors of *The Adapted Mind* and Steven Pinker, come down especially hard on the idea that the mind is a blank slate. This idea is first developed in the writings of the seventeenth-century philosopher John Locke, and is—at least according to Pinker—at the heart of something called the Standard Social Science Model. The SSSM (as it is referred to by Pinker and others) apparently grounded the social sciences throughout the twentieth century. But it must now be debunked as myth. Pinker sets out to do this in a number of his works, notably the Tanner Lecture he delivered in 1999 at Yale University. The Tanner Lectures on Human Values, as they are formally known, were established in 1978 by the American scholar and philanthropist Obert Clark Tanner with the purpose of providing the opportunity for distinguished scholars in various fields to connect their particular work to the larger study of human values. Pinker's lecture is entitled "The Blank Slate, the Noble Savage, and the Ghost in the Machine" (reprinted in *The Tanner Lectures on Human Values*, vol. 21, ed. Grethe B. Peterson [Salt Lake City: University of Utah Press, 2000], 181–209). In it he attempts quite directly to refute what he feels are the three false beliefs that support the SSSM. These beliefs are, roughly: the idea that the mind is a blank slate; the idea that humans in the "state of nature" are "noble" and "good"; and the idea that the essence of human consciousness resides in an immaterial substance within us. As the book you are reading goes to press, Pinker's latest book, *The Blank Slate: The Modern Denial of Human Nature* (New York: Viking Press, 2002), has just been published. Pinker's 2002 book expands on the arguments he presents in his Tanner lecture and makes an important turn in the direction of exploring the impact that the theories put

forth by evolutionary psychologists are having on our public discussions of who we are and how we should live.

5. The set of all rational numbers between 0 and 1 would be an example of a bounded infinite set.

6. The reason is that one can always add an additional clause to any sentence. Consider the sentence "The cat is on the mat." To construct another grammatically correct English sentence, one could simply write, "You just read the sentence, 'The cat is on the mat.' " To construct yet another grammatically correct English sentence, one could write, "I am sure that you just read the sentence, 'The cat is on the mat.' " One could continue this process infinitely, always constructing grammatically correct English sentences. See Steven Pinker, *The Language Instinct: How the Mind Creates Language* (New York: William Morrow, 1994), esp. 86–87.

7. Some may argue that certain contemporary northern European societies have essentially abolished "husbands" and "wives," given that a statistically significant percentage of young men and women in these societies seem willing and even determined to live with one another outside the bonds of marriage, and even to raise a "family." But here we really do have a case of semantics over substance. Whatever some cultural relativists with advanced degrees might like to think, an arrangement in which a man and a woman live together and enjoy more or less exclusive sexual access to each other is *essentially* a marriage. In the poorer rural parts of America, such arrangements, when not formalized, are referred to as "common law marriages." The fact that certain individuals in certain "advanced" European cultures seem not to want to use the word *marriage*—and I suspect there are more men than women who are behind this refusal—may say something about cultural fashions and tastes. But it says nothing about fundamental human nature. A more substantive refutation of the point I am making *might* come from the cultural practices of the Na, a tiny tribe in the Yunnan province of southern China. The *Western*-trained Chinese anthropologist Cai Hua made the Na the subject of his book *A Society without Fathers or Husbands: The Na of China* (translated from the French by Asti Hustvedt [New York: Zone Books, 2001]). Hua spent several years living with the Na and learning their language. He reports that, in this culture, sexual intercourse takes place when a woman "invites" a man to "visit" her where she lives—the house she usually shares with some of her sisters, brothers, and her mother's relatives. The man usually arrives in the middle of the night, stays for a period of hours, has sexual intercourse with the woman, and usually departs before dawn—without giving any further thought to the possible procreative consequences of his visit. Na women may be visited by several men on a given night. Certainly this is an unusual arrangement. Indeed, the well-known cultural anthropologist Clifford Geertz suggests that this particular aspect of Na culture "sounds like a hippie dream or a Falwell nightmare" (see "The Visit," *New York Review of Books,* 18 October 2001, 29). If the woman does become pregnant as a result of one of these

visits, the child is raised by the woman and her brothers—the child's maternal uncles. There is no requirement that a man care for his biological offspring. Indeed, unless there is a clear physical resemblance, a man may not even know his biological children. Nor, of course, would a child know his or her biological father. As Hua notes, this does raise the possibility of accidental father-daughter incest. If the reports of this cultural practice hold up, one must wonder what this does to my claim that one would never find a viable society without husbands and wives. I would begin by noting that this is certainly an area for further research. The relationship between Na culture and mental modules is not in "reflective equilibrium," as I describe that concept in chapter nine. But I would also note, as do Hua and Geertz, that to some extent this aspect of Na culture is changing, as the culture comes into contact with broader Chinese culture. But the change is not entirely the result of force or coercion. Rather, there is at least some evidence that when Na children become "aware," from their interaction with children of other cultures, that they "lack" a father, they find this troubling. Na women may also find that the support provided by biological fathers may be greater than the support they may receive from their brothers or their other maternal relatives. But, again, more research is needed.

8. W. E. B. Du Bois, *The Souls of Black Folk* (New York: Modern Library, 1996), 109–10.

9. My remarks here are directed at those in cultural studies and science studies who seek to apply postmodern or poststructuralist readings to scientific texts, in a misguided effort to "deconstruct" the *science* therein. These are the same folks whom Paul Gross and Norman Levitt critique in their book *Higher Superstition: The Academic Left and Its Quarrels with Science* (Baltimore: Johns Hopkins University Press, 1998). I am not attacking those who approach scientific texts to see how they function as instances of persuasion. Indeed, within the humanities and social sciences there is something of a renewed interest in understanding the ways in which scientific discourse—and also discourse in other fields like law and economics—is *rhetorically* constructed. Much of this interest is concerned with "unmasking" the hidden biases of these allegedly "positive" studies, so as to give academics at least a better understanding of how this discourse actually achieves its persuasive power. Some of this work is also accessible to, if not necessarily intended for, a nonacademic audience. If one wanted to trace the development of this renewed interest in the rhetoric of science (and economics and law) one could certainly do worse than begin in the early 1980s at the University of Iowa. During that time, a number of professors from various disciplines—including political science, economics, and especially communication studies—came together to form what has come to be known as the Project on the Rhetoric of Inquiry (or POROI, for short). POROI concerns itself with understanding the rhetorical dimensions of research and scholarship emanating from disciplines (like science and economics) that have not tradi-

tionally been thought of as disciplines with which rhetorical critics should concern themselves. Over the years, several hundred academics and researchers have contributed to the development of POROI, and it has also spawned several important book series. One is a series on the "Rhetoric of the Human Sciences," based at the University of Wisconsin Press, and edited by John Lyne (of the discipline of communication studies), Deirdre N. McCloskey (formerly Donald McCloskey, from the discipline of economics), and John S. Nelson (of the discipline of political science). This series includes a volume of essays edited by Jack Selzer and entitled *Understanding Scientific Prose* (Madison: University of Wisconsin Press, 1993). These essays all focus on one especially famous scientific "text"—Stephen Jay Gould and Richard Lewontin's 1979 paper "The Spandrels of San Marco and the Panglossian Paradigm: A Critique of the Adaptationist Programme." This is indeed a critically important paper in the field of evolutionary biology, and hence I discuss it at some length in chapter seven. It is also a consciously rhetorical document. Gould and Lewontin use an interesting extended metaphor drawn from religious iconography in order to persuade their colleagues to abandon strict adherence to what they call the adaptationist paradigm. *Understanding Scientific Prose* is an examination of this text from various critical perspectives. As Selzer notes, all of the contributors to the volume "aim to further the effort of rhetorical criticism that has so much momentum today. By introducing methods of analysis implied by cultural studies and the concept of intertextuality, feminism and the sociology of science, structuralism and deconstruction, reader-response theory and historicism, Perelman and Burke and Habermas (among others), we mean to extend the range of analytical approaches that are available to any critic of any discourse" (3–4).

Some of the essays in *Understanding Scientific Prose* do provide genuine insights into Gould and Lewontin's paper, and thus could have the effect of sharpening public discourse about evolution. Other essays are much less valuable in this regard. But the volume as a whole is certainly representative of the renewed interest in understanding the way scientific discourse is rhetorically constructed. For a more detailed discussion of the Project on the Rhetoric of Inquiry, see John S. Nelson, Allan Megill, and Donald McCloskey, eds., *The Rhetoric of the Human Sciences: Language and Argument in Scholarship and Public Affairs* (Madison: University of Wisconsin Press, 1991).

10. Alan D. Sokal, "Transgressing the Boundaries: Toward a Transformative Hermeneutics of Quantum Gravity," *Social Text* 46/47, nos. 1 and 2 (Spring/Summer, 1996): 217–52.

11. Derrida's comment was made at an important symposium on *Les Langages critiques et les sciences de l'homme* that was held in the 1970s and that brought together famous American and Continental philosophers and social critics, particularly those involved in the then emerging field of poststructuralist literary criticism. At that symposium a young Jacques Derrida presented a talk—later reprinted

in article form under the title "Structure, Sign, and Play in the Discourse of the Human Sciences" (in *The Structuralist Controversy: The Languages of Criticism and the Sciences of Man*, ed. Richard Macksey and Eugenio Donato [Baltimore: Johns Hopkins University Press, 1972], 247–72)—in which he sought (in what is now classical poststructuralist literary style) to, among other things, "deconstruct" the opposition between nature and culture that is so central to the humanities, or the so-called human sciences. Derrida does this by attempting to show that the structure upon which this opposition is built itself has no "center" that might function as an organizing principle of the structure. But a structure with no center, with no organization, must collapse. It must "deconstruct" itself. At the end of his talk, Derrida seems to generalize this point beyond the humanities to the entire range of Western thought and philosophy. All of it deconstructs itself. Derrida concludes that we are thus left with "two interpretations of interpretation, of structure, of sign, of freeplay" (264). The one interpretation—the interpretation that has been at the center of Western metaphysics and Western science from the beginning—"seeks to decipher, dreams of deciphering, a truth or an origin which is free from freeplay" (ibid.). This interpretation carries itself on in the name of "man"—"that being who, throughout the history of metaphysics or of ontotheology . . . has dreamed of full presence, the reassuring foundation, the origin and the end of the game" (264–65). The other interpretation, according to Derrida, is "no longer turned toward the origin." Rather, it "affirms freeplay and tries to pass beyond man and humanism" (264).

Crucially, at the end of his talk, Derrida opens the floor to questions from his colleagues at the symposium. (That question and answer exchange is itself reprinted as an appendix to Derrida's article.) The first question comes from Jean Hyppolite, who asks Derrida to elaborate on the "concept of the center of structure" (see Jean Hyppolite, "Discussion," in *The Structuralist Controversy*, 265). This request for elaboration itself comes in the context of a larger question that actually runs more than a page in length. Sokal rightly draws our attention to this particular exchange, because it goes beyond Derrida's discussion of the human sciences and focuses more on the relationship of "the game," "freeplay," and deconstruction itself *to the natural sciences*. In this sense, Sokal, in highlighting this exchange as representative of the relationship between postmodern (or at least poststructuralist) theory and the natural sciences, is being perfectly faithful to the thinking of individuals like Derrida and Hyppolite—at least to the extent that what they say is representative of what they think. In fact, at the beginning of his question, Hyppolite notes: "My question is, I think, relevant since one cannot think of the structure without the center, and the center itself is 'destructured,' is it not?—the center is not structured. I think we have a great deal to learn as we study the sciences of man; we have much to learn from the natural sciences. They are like an image of the problems which we, in turn, put to ourselves" (*The Structuralist Controversy*, 266). In his own article, Sokal actually quotes about two paragraphs of Hyppolite's question in order to demonstrate that at least a significant part of that question involves the relationship between poststructuralist theory and the natural sciences. Here, then, is part of Sokal's quotation of Hyppolite's question to Derrida regarding the concept of structure and center as it relates to the natural sciences.

> With Einstein, for example, we see the end of a kind of privilege of empiric evidence. And in that connection we see a constant appear, a constant which is a combination of space-time, which does not belong to any of the experimenters who live the experience, but which, in a way, dominates the whole construct; and this notion of the constant—is this the center? (Sokal, "Transgressing the Boundaries," 221; for the entire question by Hyppolite see "Discussion," 265–67.)

Derrida's comment on the Einsteinian constant, a portion of which I have already quoted in the text, actually goes on for some time. Three sentences, from the portion I quoted, read as follows:

> The Einsteinian constant is not a constant, is not a center. It is the very concept of variability—it is, finally, the concept of the game. In other words, it is not the concept of some*thing*—of a center starting from which an observer could master the field—but the very concept of the game. (See "Discussion.")

12. Sokal, "Transgressing the Boundaries," 222.

13. See Alan D. Sokal, "Transgressing the Boundaries: An Afterword," *Dissent*, Fall, 1996, 93–99, and Sokal, "A Physicist Experiments with Cultural Studies," *Lingua Franca*, May/June, 1996, 62–64.

14. For an interesting discussion of the development and fate of the IQ Cap, see Tom Wolfe, "Sorry, Your Soul Just Died," *Forbes ASAP*, 2 December 1996, 213–14. For a discussion of the actual science involved in correlating brainwave activity and intelligence, see Duilio Giannitrapani, *The Electrophysiology of Intellectual Functions* (New York: Karger, 1985). The Wechsler Adult Intelligence Scale is actually comprised of eleven subtests, which include, for example: "arithmetic" (the ability to solve mathematical word-problems); "vocabulary" (the ability to define words correctly); "similarities" (the ability to identify the "common elements" between two terms); "digit span" (the ability to repeat a series of digits backwards and forwards), and so forth. Interestingly, although all eleven subtests correlate very highly with brainwave activity, Giannitrapani's research seems to indicate that the subtest that correlates most highly with measurable brainwave activity—i.e., the subtest score that is best predicted by devices like the IQ Cap—is the subtest of "comprehension." According to Giannitrapani, this subtest "requires practical information and an ability to evaluate appropriate responses to situations . . . . It seems that one of the unique features of Comprehension is not only the ability to elicit responses from

within but more important the ability to evaluate the appropriateness of each elicited response, i.e. to form a judgment regarding which response is appropriate or relevant within the given context and at what point to terminate the search for a response. It could be called a test of contextual appropriateness. It does not correlate as high with Verbal IQ as do the Similarities and Vocabulary subtests" (92).

15. See Wolfe, "Sorry, Your Soul Just Died," 214, emphasis in original.

Part I. The Evolution of a Controversy

Epigraph: L. L. Thurstone, "Theories of Intelligence," *Scientific Monthly,* February 1946, 111.

Chapter 1. Stephen Jay Gould Historicizes Science

1. The title of Pinker's book *How the Mind Works* makes the point. In that book Pinker writes: "The mind is not the brain but what the brain does, and not even everything it does, such as metabolizing fat and giving off heat. . . . The brain's special status comes from a special thing the brains does, which makes us see, think, feel, choose, and act. That special thing is information processing, or computation" (see Steven Pinker, *How the Mind Works* [New York: W. W. Norton, 1997], 24).

2. Quoted in Stephen Jay Gould, *The Mismeasure of Man* (New York: W. W. Norton, 1996), 83.

3. Quoted in ibid., 69.

4. See ibid., 139, and throughout.

5. I think this is a fair summary of Herrnstein and Murray's claims. Indeed, I have chosen to include in my summary the qualifications and caveats that the authors themselves stress. See Richard J. Herrnstein and Charles Murray, *The Bell Curve: Intelligence and Class Structure in American Life* (New York: Free Press, 1996).

6. Ibid., v.

7. Gould, *The Mismeasure of Man,* 99, emphasis in original.

8. Ibid., 86.

9. Ibid., 100.

10. Ibid., 101.

11. Ibid., 98 n. 2.

12. Ibid., 126.

13. Ibid., 106, emphasis added.

14. Ibid., 54.

15. The "full history" that I do not intend to provide is, in fact, provided by Ullica Segerstråle in her excellent book *Defenders of the Truth: The Battle for Science in the Sociobiology Debate and Beyond* (Oxford: Oxford University Press, 2000).

Chapter 2. Richard Herrnstein Stirs Up Controversy at Harvard Yard

1. See Richard Herrnstein, "I.Q.," *Atlantic Monthly,* September 1971, 43–64.

2. Ullica Segerstråle, *Defenders of the Truth: The Battle for Science in the Socio-biology Debate and Beyond* (Oxford: University Press, 2000), 16.

3. For an excellent biography of Galton and his relationship to the eugenics movement, see Nicholas Wright Gillham, *A Life of Sir Francis Galton: From African Exploration to the Birth of Eugenics* (Oxford: Oxford University Press, 2001).

4. Herrnstein, "I.Q.," 47.

5. Ibid., 48, emphasis added.

6. Ibid., 48.

7. See Stephen Jay Gould, *The Mismeasure of Man* (New York: W. W. Norton, 1996), 286.

8. Herrnstein, "I.Q.," 49.

9. See Richard J. Herrnstein and Charles Murray, *The Bell Curve: Intelligence and Class Structure in American Life* (New York: Free Press, 1996), 3.

10. Ibid., 283.

11. Ibid., 3.

12. Gould uses the example of Halley's comet in the updated, 1996, version of his book, although he does not relate its distance from the earth to the growth of websites on the world wide web. See Gould, *The Mismeasure of Man,* 272.

13. Of course, the distance of Halley's comet from the earth has been increasing at a fairly steady rate, while the number of websites on the world wide web has been increasing in a somewhat more complicated fashion.

14. Gould, *The Mismeasure of Man,* 269.

15. See ibid., 292.

16. An important qualification is that the actual population must of course have a finite variance.

17. William Youden, quoted in Thomas H. Wonnacott and Ronald J. Wonnacott, *Introductory Statistics* (3d ed.; New York: John Wiley and Sons, 1977), 143.

18. Galton, quoted in ibid.

19. Charles Murray, "Afterword" to *The Bell Curve,* 561.

20. Ibid., emphasis added.

21. See "Mainstream Science on Intelligence," *Wall Street Journal,* 13 December 1994, A 18.

22. Murray, "Afterword" to *The Bell Curve,* 562.

23. See Arthur Jensen, "How Much Can We Boost IQ and Scholastic Achievement?" *Harvard Educational Review* 39 (1969): 1–123.

24. Jensen, quoted in Herrnstein, "I.Q.," 54.

25. Ibid.

26. Herrnstein and Murray, *The Bell Curve,* 9.

27. Herrnstein, "I.Q.," 55.

28. Ibid.

29. Ibid.

30. Ibid., 57.

31. Ibid., 63

32. Ibid., 64, emphasis in original.

33. See Nicholas Lemann, *The Big Test: The Secret History of the American Meritocracy* (New York: Farrar, Straus, and Giroux, 1999), esp. 27–52.

34. Ibid., 116.

35. Economists will recognize the society thus described as one that is Pareto optimal. A Pareto optimal arrangement is any arrangement in which no individual can be made better off without at least one individual being made worse off. If economic arrangements in a society are not Pareto optimal, then there must exist an arrangement in which one individual can be made better off without a loss to anyone. As Rawls and others note, this is obviously a first requirement for almost any society or any economic arrangement. See John Rawls, *A Theory of Justice* (Cambridge, Mass.: Harvard University Press, 1971), esp. 67–75.

36. Ibid., esp. 76–83.

37. Rawls's inability to provide a convincing argument on this point, and the serious philosophical damage that lack does to his overall theory, is most thoroughly discussed by Robert Nozick in his *Anarchy, State, and Utopia* (New York: Basic Books, 1977).

38. See Clark Kerr, *The Uses of the University* (Cambridge, Mass.: Harvard University Press, 1982).

39. Ibid., 121.

40. Again, this is the conflict between elitism and egalitarianism—a conflict that is particularly acute at elitist, liberal institutions like Harvard. Indeed, as Allan Bloom (that old foe of the cultural left) notes with evident relish, if those at Harvard who speak so loudly of the value of egalitarianism really took their ideas seriously, the first and most obvious course of action would be to press hard for open, random admissions to Harvard. The fact that you do not see many Harvard faculty members advocating such a change suggests to Bloom at least that these faculty members may believe more than they care to say that a Harvard education is too valuable to waste on just anyone. See Allan Bloom, "Western Civ," in *Giants and Dwarfs* (New York: Simon and Schuster, 1990), esp. 14–18. But there may actually be some members of Harvard's faculty who do support something akin to "open admissions." Robert Bork claims that Duncan Kennedy, a Harvard law professor well known in the Critical Legal Studies movement, advocates that "admission to law school should be by lot, with quotas within the lottery for minority, female, and

working-class students. The purpose of the lottery device is to eliminate notions of merit and of better and poorer law schools." Since quotas would not be needed to insure *equal* representation of any subgroup in law school, assuming that the process of admission were strictly random, one must conclude that Kennedy's quotas for minority, female, and working-class students are designed to insure that they are *overrepresented* in laws schools when compared to their percentages in the overall population. Presumably this is done to hasten the day when the percentages of working lawyers who are minorities, females, and the children of working-class individuals will be equal to the percentages of those groups in the population at large. For a further discussion of this see Robert Bork, *The Tempting of America: The Political Seduction of the Law* (New York: Free Press, 1990), 208.

41. See Richard Herrnstein, *IQ in the Meritocracy* (Boston: Atlantic–Little Brown, 1973).

Chapter 3. Edward O. Wilson Brings More Controversy to the Yard

1. See Edward O. Wilson, *Sociobiology: The New Synthesis* (Cambridge, Mass: Harvard University Press, 1975), x.

2. Ibid., 4.

3. Edward O. Wilson, *On Human Nature* (Cambridge, Mass.: Harvard University Press, 1978), 16.

4. Wilson, *Sociobiology*, 562.

5. Ibid.

6. See ibid., 551, 553–54, and 565.

7. Ibid., 555.

8. Ibid.

9. Ibid., 550.

10. Segerstråle notes that "a careful reading of *Sociobiology* shows that Wilson was extremely cautious when it came to matters of intelligence; in fact, he played down the social significance of IQ-type intelligence. . . . And not only did he not make overtly racist statements, but he also approvingly cited the modern position that 'race' is not a meaningful biological concept." See Ullica Segerstråle, *Defenders of the Truth: The Battle for Science in the Sociobiology Debate and Beyond* (Oxford: Oxford University Press, 2000), 27.

11. See ibid., esp. 35–51.

12. Segerstråle makes a similar point. She writes, "Wilson appears to have actively *sought out* particular persons to serve the role as immediate competitors—before Lewontin, it was Jim Watson [the co-discoverer of the structure of the DNA molecule and a professor of biology at Harvard] that was Wilson's stimulating challenge." Ibid., 49, emphasis in original.

13. For a discussion of Chomsky's problems with the Sociobiology Study Group see Segerstråle, *Defenders of the Truth*, esp. 204–5.

14. See Elizabeth Allen, et al., "Against Sociobiology," *New York Review of Books*, 13 November 1975, 43–44.

15. Ibid., 45.

16. Ibid.

17. Ibid.

18. Ibid., 43.

19. Ibid.

20. See Edward O. Wilson, "For Sociobiology," *New York Review of Books*, 11 December 1975, 60–61.

21. Segerstråle, *Defenders of the Truth*, 23.

22. Ibid., 20.

23. See Edward O. Wilson, *Naturalist* (Washington, D.C.: Island Press, 1994).

24. Tom Wolfe, "Sorry, But Your Soul Just Died," *Forbes ASAP*, 2 December 1996, 212.

25. Segerstråle, *Defenders of the Truth*, 45.

26. Wilson, *Sociobiology*, 575.

27. John F. Kennedy, "Commencement Address at Yale University, 1962," quoted in Allen J. Matusow, *The Unraveling of America: A History of Liberalism in the 1960s* (New York: Harper and Row, 1984), 48.

28. Wilson, *On Human Nature*, 133.

29. Ibid.

30. See Segerstråle, *Defenders of the Truth*, 212.

31. In the early 1990s I had the opportunity to discuss the practice of gender norming at some length with cadets (both male and female) and their instructors at the United States Military Academy. Two things struck me most from these extensive, though obviously unscientific, discussions. First, everyone seemed to have made his or her peace with the practice of gender norming. Both males and females reiterated the line that gender norming was just something that they needed to "deal with." Second, and much more importantly, gender norming has seemed to signal to cadets (both male and female, and especially those not particularly physically gifted) that physical training requirements do not really (or *should* not really) matter all that much. When a male and a female cadet receive the same "adjusted" score for significantly different levels of performance on a physical training test, all cadets quickly come to understand that their teachers cannot really be serious about this part of the cadet experience. I repeatedly heard the opinion voiced that the Academy ought simply to drop most of its physical training requirements, or at least make most training exercises optional, so that students could concentrate on what was really important—i.e., their studies.

32. Wilson, *On Human Nature,* 132.

33. Edward O. Wilson, *Consilence: The Unity of Knowledge* (New York: Knopf, 1998), 277.

34. See, for example, Edward O. Wilson, *Biophilia* (Cambridge, Mass.: Harvard University Press, 1984), and Wilson, *The Future of Life* (New York: Knopf, 2002).

35. Wilson, *Consilence,* 3.

36. See Segerstråle, *Defenders of the Truth,* 263.

37. For a fuller discussion of Salvador Luria's connection to the sociobiology debate, see ibid., 245–53.

38. See Wilson, *Naturalist,* 339.

39. See Segerstråle, *Defenders of the Truth,* 39.

40. Ibid., 262, footnote omitted.

41. Wilson, *Sociobiology,* 3.

42. Although, having said that, Wilson does come down fairly hard on some religious strictures—notably the ban, by some religions, on artificial methods of birth control. Wilson criticizes this prohibition on the grounds that it is a complete misapplication of "natural law." He is probably correct here. But that is at least arguable. Oddly, Wilson is silent on the best argument against this prohibition, which is that it is so hopelessly confused about what counts as *artificial.* It was G. K. Chesterton (I think) who put this point best when he noted that the Catholic Church (for example) allows you to use mathematics, but not physics or chemistry, to practice birth control.

43. Wilson, *Consilence,* 277.

44. Karl Marx, "The German Ideology," in *Karl Marx: Selected Writings,* ed. David McLellan (Oxford: Oxford University Press, 1977), 159.

45. For a discussion of Marx's various rhetorical problems, see James Arnt Aune, *Rhetoric and Marxism* (Boulder, Colo.: Westview Press, 1994).

46. Marx, "The German Ideology," 169.

47. Wilson, *Consilence,* 246.

Chapter 4. Richard Lewontin and His Colleagues Demur

1. In 1984, Ronald Reagan received 525 electoral votes, and his opponent, Walter Mondale, received 13. But Reagan's margin of victory was actually less than Franklin Roosevelt's margin of victory in 1936. Then, during the height of the Depression, Roosevelt received 523 votes, while his opponent, Alfred Landon, received 8 votes.

2. See Mario Cuomo, "Keynote Address at the Democratic National Convention, San Francisco, California, July 17, 1984," in *Vital Speeches of the Day,* 15 August 1984, 646. Cuomo develops the theme of Social Darwinism throughout his speech, concluding with these words about his party: "We believe in encouraging the talented, but we believe that while survival of the fittest may be a good working

description of the process of evolution, a government of humans should elevate itself to a higher order, one which fills the gaps left by chance or a wisdom we don't understand. We would rather have laws written by the patron of this great city, the man called the 'world's most sincere Democrat,' St. Francis of Assisi, than laws written by Darwin" (648). Republicans must have been listening to Cuomo's persuasive linkage of their party's ideology with Social Darwinism. It is no coincidence that in 2000 George W. Bush called himself a "compassionate conservative."

3. See Richard Lewontin, Steven Rose, and Leon Kamin, *Not in Our Genes: Biology, Ideology, and Human Nature* (New York: Pantheon Press, 1984).

4. See *Alas, Poor Darwin: Arguments against Evolutionary Psychology,* ed. Hilary Rose and Steven Rose (New York: Harmony Books, 2000).

5. See Leon Kamin, *The Science and Politics of IQ* (New York: Halsted Press, 1974).

6. Lewontin et al., *Not in Our Genes,* 11–12.

7. Some of the evidence for the heritability of religiosity comes from studies of monozygotic twins reared apart. I discuss one such study, by Thomas Bouchard and others, more below in the context of a discussion of the heritability of intelligence. See Thomas J. Bouchard Jr., David T. Lykken, Matthew McGue, Nancy L. Segal, and Auke Tellegen, "Sources of Human Psychological Difference: The Minnesota Study of Twins Reared Apart," *Science,* 12 October 1990, 223–28.

8. Obviously the point I am making by using this example is a theoretical one. While I assume few people would disagree with the claim that, at most colleges, the environmental influences on a given student vary considerably, I will readily concede that it is quite a bit more difficult to know how much less the variation with respect to environmental influences may be for students at VMI. For an engaging discussion of the VMI experience, see Laura Fairchild Brodie, *Breaking Out: VMI and the Coming of Women* (New York: Pantheon Press, 2000).

9. For a discussion of this point, see Richard J. Herrnstein and Charles Murray, *The Bell Curve: Intelligence and Class Structure in American Life* (New York: Free Press, 1996), 111–13.

10. See Lewontin et al., *Not in Our Genes,* 83–129.

11. Ibid., 112.

12. See ibid., 111–12.

13. Ibid., 113.

14. Ibid., emphasis in original.

15. See Thomas J. Bouchard Jr. et al., "Sources of Human Psychological Difference: The Minnesota Study of Twins Reared Apart," 227.

16. Ibid., 224.

17. See ibid., 224–25.

18. Ibid., 226, emphasis added. Actually, the ratios of the correlations on several measures of intelligence between monozygotic twins reared apart (MZA) and

monozygotic twins reared together (MZT), while not as high as those for most of the ratios concerning physiological traits, were still in the ballpark. For example, the ratio for height was 0.925, while the ratio for IQ as measured by the Wechsler Adult Intelligence Scale was 0.784.

19. Ibid., 227.

20. Lewontin et al., *Not in Our Genes,* 116.

21. Herrnstein and Murray, *The Bell Curve,* 108, footnote omitted.

22. Lewontin et al., *Not in Our Genes,* 136.

23. Ibid., 239.

24. Ibid., 267.

25. Ibid., 268.

26. Ibid.

27. Ibid., 270.

28. Richard Dawkins, "Sociobiology: The Debate Continues," *New Scientist,* 24 January 1985, 60.

29. Lewontin et al., *Not in Our Genes,* 282.

30. Patrick Bateson, "Sociobiology: The Debate Continues," *New Scientist,* 24 January 1985, 59.

31. Lewontin et al., *Not in Our Genes,* 289–90.

Part II. The Blind Watchmaker Meets the Scatterbrained Computer Programmer

Epigraph: Steven Pinker, *How the Mind Works* (New York: W. W. Norton, 1997), 23, emphasis in original.

Chapter 5. Nature's "Very Special Way"

1. Richard Dawkins, *The Blind Watchmaker: Why the Evidence of Evolution Reveals a Universe without Design* (New York: W. W. Norton, 1996).

2. Quoted in Dawkins, *The Blind Watchmaker,* 5.

3. Immanuel Kant, *Critique of Judgment,* trans. Werner S. Pluhar (Indianapolis: Hackett, 1987), 253, emphasis added.

4. Ibid., 254.

5. Ibid., 256, emphasis in original.

6. Ibid.

7. Dawkins, *The Blind Watchmaker,* 5.

8. Ibid., emphasis in original.

9. See Geoffrey Miller, *The Mating Mind: How Sexual Choice Shaped the Evolution of Human Nature* (New York: Anchor Books, 2000), 47.

10. See Christopher Badcock, *Evolutionary Psychology: A Critical Introduction* (Cambridge: Polity Press, 2000), 155–56.

11. For the entire list, see Biography of the Millennium Names the Top 100. (n.d.). Retrieved 23 May 2001, from the world wide web: http://www.biography .com/features/millennium.

12. See Steve Jones, *Darwin's Ghost: The Origin of Species Updated* (New York: Random House, 1999), xix.

13. See Charles Darwin, *The Origin of Species by Means of Natural Selection or The Preservation of Favored Races in the Struggle for Life* (New York: Modern Library, 1993). "Variation under Domestication" is the first chapter of the book; see pages 24–64. For a discussion of the rhetorical aspects of Darwin's argument in *The Origin of Species*, see John A. Campbell, "Scientific Discovery and Rhetorical Invention: The Path to Darwin's Origin," in Herbert W. Simon, ed., *The Rhetorical Turn: Invention and Persuasion in the Conduct of Inquiry* (Chicago: University of Chicago Press, 1990), 58–90.

14. Aristotle, *Rhetoric*, trans. John Henry Freese, Loeb Classical Library (Cambridge, Mass.: Harvard University Press, 1982), 467.

15. Darwin, *The Origin of Species*, 108.

16. See ibid., esp. 331. Richard Dawkins also uses this example in *The Selfish Gene* (New York: Oxford University Press, 1989), esp. 133.

17. Darwin, *The Origin of Species*, 331.

18. For a truly fascinating discussion of the human ability to perceive colors, see Roger Shepard, "The Perceptual Organization of Colors: An Adaptation to Regularities of the Terrestrial World," in Jerome H. Barkow, Leda Cosmides, and John Tooby, eds., *The Adapted Mind: Evolutionary Psychology and the Generation of Culture* (Oxford: Oxford University Press, 1992), 495–532.

19. John Tooby and Leda Cosmides, "The Psychological Foundations of Culture," in Barkow, Cosmides, and Tooby, eds., *The Adapted Mind*, 56.

20. For a discussion of the design of the eye, see Dawkins, *The Blind Watchmaker*, 93.

21. Darwin, *The Origin of Species*, 231, emphasis added.

22. See Stephen Jay Gould, "The Tallest Tale," in *Leonardo's Mountain of Clams and the Diet of Worms: Essays on Natural History* (New York: Harmony Books, 1998), 301–18.

23. See Jones, *Darwin's Ghost*, 1–20.

24. See Miller, *The Mating Mind*, esp. 230–37. In these pages Miller discusses the evolution of the human penis and explains how its various features almost certainly resulted from female choice acting as a selection pressure.

25. As Sarah Hrdy points out, "Such male [langur monkeys] are often in a special state of arousal [when on these killing sprees], as evidenced by an erect penis,

though without other indications of sexual interest." See Sarah Blaffer Hrdy, *Mother Nature: A History of Mothers, Infants, and Natural Selection* (New York: Pantheon Press, 1999), 237.

26. Stephen Jay Gould makes this point about Darwin's championing of the unpredictability of life. See the preface to the revised edition of Stephen Jay Gould, *Questioning the Millennium: A Rationalist's Guide to a Precisely Arbitrary Countdown* (New York: Harmony Books, 1999), 32–33.

27. Darwin, *The Origin of Species*, 648–49.

Chapter 6. What Is the Mind?

1. See Richard Dawkins, *The Blind Watchmaker: Why the Evidence of Evolution Reveals a Universe without Design* (New York: W. W. Norton, 1996), 27–28.

2. The apparently unique capacity for self-awareness is what gave rise to the whole "mind-body" problem in philosophy. In a sense, this is the problem of "consciousness," but of the particularly human consciousness of which a being with a language that can refer to itself is capable. Some philosophers claim that the very self-referential character of human thought—our ability to think that we are thinking—means that thinking cannot be a material process; hence the mind cannot be a material substance. For an engaging discussion of this point, see Roger Penrose, *The Emperor's New Mind: Concerning Computers, Minds, and the Laws of Physics* (Oxford: Oxford University Press, 1989).

3. On this point, Immanuel Kant gave the romantic poets their best "philosophical" gift of justification, by arguing that "nature," working by the means of some fundamentally inexplicable process, confers upon the artist alone originality and creativity. Kant then links originality and creativity to genius, and concludes that only the artist can be a genius. Scientists like Newton, however intelligent they may have been, do not, according to Kant, possess the quality of genius because their manner of thinking is not original. It is, in a precise sense, derivative, insofar as we can derive all of the ideas in the *Principia* in a logical manner. But one cannot derive the ideas in, say, Milton's poetry logically. Kant would surely have agreed that the essential quality of the mind that could never be duplicated by a computer, or any other material process, is the mind's creativity. Consider this passage on genius and originality from Kant's *Critique of Judgment*, trans. Werner S. Pluhar (Indianapolis, Ind.: Hackett, 1987), 175.

> Genius is a talent for producing something for which no determinate rule can be given, not a predisposition consisting of a skill for something that can be learned by following some rule or other; hence the foremost property of genius must be originality. . . . Genius itself cannot describe or indicate scientifically how it brings about it products, and it is rather as nature that it gives

the rule. That is why, if an author owes a product to his genius, he himself does not know how he came by the ideas for it; nor is it in his power to devise such products at his pleasure, or by following a plan, and to communicate [his procedure] to others in precepts that would enable them to bring about like products.

4. Pinker writes: "This is computation, I claim, but that does not mean that the computer is a good metaphor for the mind." See Steven Pinker, *How the Mind Works* (New York: W. W. Norton, 1997), 23.

5. Ibid., 36.

6. Jerome H. Barkow, Leda Cosmides, and John Tooby, "Introduction: Evolutionary Psychology and Conceptual Integration," in Jerome H. Barkow, Leda Cosmides, and John Tooby, eds., *The Adapted Mind: Evolutionary Psychology and the Generation of Culture* (Oxford: Oxford University Press, 1992), 7.

7. Ibid., 8, emphasis in original.

8. Ibid.

9. Pinker, *How the Mind Works*, 21.

10. Ibid., 31.

11. Ibid., 27.

12. Ibid., 131.

13. Ibid., 85.

14. Ibid.

15. See Sarah Blaffer Hrdy, *Mother Nature: A History of Mothers, Infants, and Natural Selection* (New York: Pantheon Press, 1999), especially 288–317.

16. For a discussion of contemporary college psychology textbooks, see Steven Pinker, *The Language Instinct: How the Mind Creates Language* (New York: William Morrow, 1994), 421.

17. See Thomas Kuhn, *The Structure of Scientific Revolutions* (Chicago: University of Chicago Press, 1962).

18. For a discussion of face recognition, see Pinker, *How the Mind Works*, 272–74.

19. For a discussion of autism and mental modules, see Pinker, *How the Mind Works*, 330–33, and Christopher Badcock, *Evolutionary Psychology: A Critical Introduction* (Cambridge: Polity Press, 2000), 113–15.

20. See Irwin Silverman and Marion Eals, "Sex Differences in Spatial Abilities: Evolutionary Theory and Data," in Barkow, Cosmides, and Tooby, eds., *The Adapted Mind* (Oxford: Oxford University Press, 1992), 533–49.

21. See Alan Turing, "Computing Machinery and Intelligence," *Mind* 59 (1950): 433–60.

22. I discuss the philosophy of the imitation game more extensively in my book *The Last Conceptual Revolution: A Critique of Richard Rorty's Political Philosophy* (Albany: SUNY Press, 1999), esp. 151–53. There I raise the provocative question,

What will happen when the interrogator has less than a fifty-fifty chance of winning the game? In other words, what will happen when a computer is better at pretending to be a human than a human is at being a human?

23. Pinker discusses the fairly dismal performance of computers when playing the imitation game, and speculates on why the game is so difficult for machines. See Pinker, *The Language Instinct*, esp. 192–96.

24. Turing, "Computing Machinery and Intelligence," 442.

Chapter 7. The Challenges of Reverse Engineering

1. See Rudyard Kipling, *Just So Stories* (New York: Puffin Books, 1994).

2. Steven Pinker, *How the Mind Works* (New York: W. W. Norton, 1997), 21.

3. Daniel C. Dennett, *Darwin's Dangerous Idea: Evolution and the Meanings of Life* (New York: Simon and Schuster, 1995), 212.

4. See ibid., 213, note 16.

5. Pinker, *How the Mind Works*, 201.

6. See E. J. Dionne, "Buchanan Ducks Comparisons to Duke," *Star Tribune* (Minneapolis, Minn.), 1 March 1992, http://web.lexis-nexis.com (14 August 2001).

7. Jerome H. Barkow, Leda Cosmides, and John Tooby, "Introduction: Evolutionary Psychology and Conceptual Integration," in Jerome H. Barkow, Leda Cosmides, and John Tooby, eds., *The Adapted Mind: Evolutionary Psychology and the Generation of Culture* (Oxford: Oxford University Press, 1992), 5, emphasis added, citations omitted.

8. Steven Pinker and Paul Bloom, "Natural Language and Natural Selection," in *The Adapted Mind,* 451.

9. Geoffrey Miller, *The Mating Mind: How Sexual Choice Shaped the Evolution of Human Nature* (New York: Anchor Books, 2000), 181.

10. Sarah Blaffer Hrdy, *Mother Nature: A History of Mothers, Infants, and Natural Selection* (New York: Pantheon Press, 1999), 145, emphasis in original.

11. See Pinker, *How the Mind Works,* 525.

12. For a discussion of this explanation for motion sickness, see Michel Treisman, "Motion Sickness: An Evolutionary Hypothesis," *Science* 197 (1977): 493–95, and Pinker, *How the Mind Works,* 265–66.

13. The paper is reprinted as an appendix in Jack Selzer, ed., *Understanding Scientific Prose* (Madison: University of Wisconsin Press, 1993).

14. See ibid.

15. See, for example, Robert Marks, "Architecture and Evolution," *American Scientist* 84 (July–August 1996): 383–89.

16. See Stephen Jay Gould, "Fulfilling the Spandrels of World and Mind," in Selzer, ed., *Understanding Scientific Prose,* 310–36.

17. John Lyne, "Angels in the Architecture: A Burkean Inventional Perspective on 'Spandrels,' " in ibid., 144–45, emphasis added.

18. Stephen Jay Gould and Richard Lewontin, "The Spandrels of San Marco and the Panglossian Paradigm: A Critique of the Adaptationist Programme," in ibid., 339–40.

19. Ibid., 340.

20. Ibid., 341.

21. Ibid., 342–3.

22. Pinker and Bloom, "Natural Language and Natural Selection," 482.

23. Miller, *The Mating Mind*, 352.

24. Ibid., 354–55.

25. Ibid., 355.

26. Ibid., 375.

27. Ibid., 376–77.

Part III. The Nature of Human Cultures

Epigraph: Jerome H. Barkow, Leda Cosmides, and John Tooby, "Introduction: Evolutionary Psychology and Conceptual Integration," in Jerome H. Barkow, Leda Cosmides, and John Tooby, eds., *The Adapted Mind: Evolutionary Psychology and the Generation of Culture* (Oxford: Oxford University Press, 1992), 3.

Chapter 8. The Benefits of Hardwiring

1. For the claim about Fischer's bizarre strategy, see Daniel Dennett, *Darwin's Dangerous Idea: Evolution and the Meanings of Life* (New York: Simon and Schuster, 1995), 213 n. 16. For an interesting discussion of the workings of Deep Blue and the games it played with Kasparov, see Monroe Newborn and Monty Newborn, *Kasparov versus Deep Blue: Computer Chess Comes of Age* (New York: Springer Verlag, 1996).

2. See "In New Casino Contest, Henhouse Has the Edge," *New York Times,* 5 September 2001. Retrieved 16 January 2002 from the world wide web: *http://www.ny times.com/2001/09/05/nyregion/05CHIC.html.* Animal rights activists were concerned that the chickens were being exploited.

3. The figure for possible forty-move chess games is from Pinker. He arrives at it by assuming that there are roughly thirty-five possible moves at any point in a chess game and that a move can (obviously) be followed by thirty-five other moves, for roughly one thousand possible moves per turn. Hence the number of moves in forty turns is one thousand to the fortieth power. See Pinker, *How the Mind Works* (New York: W. W. Norton, 1997), 137.

4. Of course, these different configurations of sets of chess pieces can themselves be combined to form an almost unimaginatively large number of full chessboard configurations. Perhaps an analogy will be helpful. Think of sets of configurations of chess pieces as words or idioms. Using rules, words or idioms can be combined to form sentences. Similarly, using rules, sets of configurations of chess pieces can be combined to form full chessboard configurations. (You will know you have violated one of the combinatorial rules for English words if you see a transitive verb—like *to make*—without a direct object. Thus "He makes" is not generally considered a grammatical combination of English words, but "He rules" is. Similarly, you will know that you have violated one of the combinatorial rules for chess configurations if you see a configuration with the opposing kings on adjacent squares.) Research seems to show that the greater one's vocabulary—the more words and idioms one can readily retrieve from memory—the better one's reading skills. But here is the really interesting point. Words and idioms are content-specific to reading skills. One who reads a good deal on evolutionary psychology and hence has a good deal of background knowledge in the subject will be better able to read another book on evolutionary psychology than will another individual who has spent the same amount of time reading not evolutionary psychology but American history. The reason is that the person schooled in evolutionary psychology does not need to spend precious seconds reminding himself exactly what *selection pressure* means every time he encounters that idiom, just as the person schooled in American history does not need to spend precious seconds reminding himself what the Missouri Compromise was every time he runs across that idiom. This all sounds painfully obvious, but it is perhaps the central insight of E. D. Hirsch's much maligned book *Cultural Literacy: What Every American Needs to Know* (New York: Vintage Books, 1988). Hirsch develops this point by drawing on empirical research and studies done on reading skills *and* on chess playing ability. Indeed, it seems that the success of the chess grandmaster compared to an average player may lie not so much in the grandmaster's superior analytical or inferential reasoning skills, or even in his general intelligence, as in his ability to recognize a very large number of sets of configurations of chess pieces. For further fascinating insights on skill in chess see Hirsch, *Cultural Literary*, 60–64.

5. See Steven Pinker, *The Language Instinct: How the Mind Creates Language* (New York: William Morrow, 1994), esp. 32–39.

6. See ibid., 323. See also Christopher Badcock, *Evolutionary Psychology: A Critical Introduction* (Cambridge, Mass.: Polity Press, 2000), 246–53.

7. See Pinker, *The Language Instinct*, esp. 55–82.

8. Quoted in Pinker, *The Language Instinct*, 59, emphasis in original.

9. Ibid., 423–24.

10. Ibid., 424; see also Pinker, *How the Mind Works*, 325–27.

11. See Badcock, *Evolutionary Psychology*, esp. 182–88.

12. Pinker, *How the Mind Works,* 35.

13. Geoffrey Miller, *The Mating Mind: How Sexual Choice Shaped the Evolution of Human Nature* (New York: Anchor Books, 2000), 14, emphasis added.

14. Ibid., 372–73.

15. Ibid., 222–23.

Chapter 9. What Cultures Can the Mind Run?

1. See Clifford Geertz, *The Interpretation of Cultures: Selected Essays* (New York: Basic Books, 1973), 5.

2. See Jerome H. Barkow, Leda Cosmides, and John Tooby, eds., *The Adapted Mind: Evolutionary Psychology and the Generation of Culture* (Oxford: Oxford University Press, 1992), and Jerome H. Barkow, *Darwin, Sex, and Status: Biological Approaches to Mind and Culture* (Toronto: University of Toronto Press, 1989).

3. See Geoffrey Miller, *The Mating Mind: How Sexual Choice Shaped the Evolution of Human Nature* (New York: Anchor Books, 2000); Bruce J. Ellis, "The Evolution of Sexual Attraction: Evaluative Mechanisms in Women," in Barkow, Cosmides, and Tooby, eds., *The Adapted Mind,* 267–88; and David Buss, "Mate Preference Mechanisms: Consequences for Partner Choice and Intrasexual Competition," in Barkow, Cosmides, and Tooby, eds., *The Adapted Mind,* 249–66.

4. See Pinker, *How the Mind Works* (New York: W. W. Norton, 1997), 470.

5. For a general discussion of this point, see Susan Faludi, *Backlash: The Undeclared War against American Women* (New York: Crown, 1991), and Naomi Wolf, *The Beauty Myth: How Images of Beauty Are Used against Women* (New York: Anchor Books, 1991).

6. For a discussion of the waist-to-hip ratio research, see Christopher Badcock, *Evolutionary Psychology: A Critical Introduction* (Cambridge, Mass.: Polity Press, 2000), 162–63; Miller, *The Mating Mind,* 248; Pinker, *How the Mind Works,* 485–87.

7. See, for example, J. M. Townsend and G. D. Levy, "Effects of Potential Partners' Physical Attractiveness and Socioeconomic Status on Sexuality and Partner Selection," *Archives of Sexual Behavior* 19 (1990): 149–64, and Ellis, "The Evolution of Sexual Attraction," 267–88.

8. See Badcock, *Evolutionary Psychology,* 161.

9. See ibid.

10. For a discussion of the evolved facial features of men and women in general, see Pinker, *How the Mind Works,* 484. For a discussion of facial attractiveness, see D. Parrett et al., "Effects of Sexual Dimorphism on Facial Attractiveness," *Nature* 394 (1998): 884–87, and I. Penton-Voak, "Menstrual Cycle Alters Face Preference," *Nature* 399 (1999): 741–42. See also Badcock, *Evolutionary Psychology,* 161.

11. Sarah Blaffer Hrdy, *Mother Nature: A History of Mothers, Infants, and Natural Selection* (New York: Pantheon Press, 1999), 226.

12. Ellis, "The Evolution of Sexual Attraction," 273, emphasis in original.

13. On sexual fantasies and differences in the deployment of jealousy, see Badcock, *Evolutionary Psychology*, 174–75; on differences in visual arousal, see Pinker, *How the Mind Works*, 471–72; on similarities, see David Buss, "Mate Preference Mechanisms," 254–56; on men, women, and children, see Ellis, "The Evolution of Sexual Attraction," 273.

14. See John Rawls, *A Theory of Justice* (Cambridge, Mass.: Harvard University Press, 1971), esp. 48–51.

15. See Derek Freeman, *Margaret Mead and Samoa: The Making and Unmaking of an Anthropological Myth* (Cambridge, Mass.: Harvard University Press, 1983).

16. Hrdy, *Mother Nature*, 229.

17. See Miller, *The Mating Mind*, esp. 122–29.

18. New York City is truly a unique place. As tourists know, you can purchase "knock-off" items on the street—like a "genuine" Louis Vuitton bag or an equally "genuine" Rolex watch—usually for about twenty dollars. Of course, when you get back to Peoria no one need know where you got your Rolex. In addition to "genuine" designer items, one can also purchase on the streets of New York little blue boxes with the Tiffany name—boxes that are almost indistinguishable from the real thing. Mind you, there is nothing *in* the boxes. But it is really the box that counts.

19. For a discussion of this issue from an evolutionary psychology perspective, see Kingsley Browne, *Divided Labours: An Evolutionary View of Women at Work* (New Haven, Conn.: Yale University Press, 1998), esp. 42–51.

20. There is even a fourth option for women. They can choose to pursue high status in a career only and thus forsake a potential mate, but at the same time choose to have children—by, for example, becoming artificially inseminated—and then to raise the children so conceived without a father.

21. Richard Dawkins, *The Selfish Gene* (Oxford: Oxford University Press, 1989), 164–65.

22. See Naomi Wolf, *The Beauty Myth*, esp. 288–89.

Chapter 10. The Evolutionary Psychology of "Little House on the Prairie"

1. Robert Trivers, "The Evolution of Reciprocal Altruism," *Quarterly Review of Biology* 46 (March 1971): 35–57.

2. Ibid., 35.

3. For a good example of such a book, see William Poundstone, *Prisoner's Dilemma: John von Neumann, Game Theory, and the Puzzle of the Bomb* (New York: Anchor Books, 1992).

4. In an iterated Prisoner's Dilemma game, which we will come to shortly, the "average" of the "payoff" from being cheated and from cheating must not exceed the payoff from mutual cooperation.

5. See Allen's comment on Plato's *Gorgias* in *The Dialogues of Plato*, vol. 1, trans. R. E. Allen (New Haven, Conn.: Yale University Press, 1984), 212–15.

6. See Robert Axelrod, *The Evolution of Cooperation* (New York: Basic Books, 1984); Robert Axelrod and William Hamilton, "The Evolution of Cooperation," *Science* 211 (1981): 1390–96; Richard Dawkins, *The Selfish Gene* (Oxford: Oxford University Press, 1989), 202–33; and Leda Cosmides and John Tooby, "Cognitive Adaptations for Social Exchange," in Jerome H. Barkow, Leda Cosmides, and John Tooby, eds., *The Adapted Mind: Evolutionary Psychology and the Generation of Culture* (Oxford: Oxford University Press, 1992), 163–228.

7. See Trivers, "The Evolution of Reciprocal Altruism," 52.

8. Ibid., 37, emphasis in original.

9. See ibid., 36–38.

10. This is, of course, the *flip side* of the view that our brains evolved as quickly and as complexly as they did so that we could take advantage of *others* (cheat) in situations of social exchange. Trivers discusses this more Machiavellian view in his article. See also Nicholas Humphrey, "The Social Function of Intellect," in P. P. G. Bateson and R. A. Hinde, eds., *Growing Pains in Ethology* (Cambridge: Cambridge University Press, 1976), 303–17; and Steven Pinker, *How the Mind Works* (New York: W. W. Norton, 1997), 193.

11. See Trivers, "The Evolution of Reciprocal Altruism," 45.

12. The examples I use are adapted from the work of Leda Cosmides and John Tooby. See their "Cognitive Adaptations for Social Exchange."

13. See ibid., esp. 187.

14. See ibid. There may be more than one module at work here, of course.

15. See ibid, esp. 188. I think that these two formulations are largely equivalent, although there may be something of a shade of difference that might be important when considering not cheating, but altruism.

16. Ibid., 188.

17. See ibid., esp. 188–89.

18. See ibid., esp. 186 and 196–97.

19. Cosmides and John Tooby's original work has spawned much discussion and much controversy. It has also generated a fairly large body of further research designed to determine precisely what is going on when we reason about various types of situations. I think that I have accurately summarized Cosmides and Tooby's work *as it applies to the argument I am trying to develop in this chapter.* But I want to emphasize two additional points for clarification.

First, all of the research in this area very strongly indicates that when subjects are explicitly cued to the possibility of cheating—or when they simply assume the possibility because their culture makes it obvious—they reason dramatically better than they do without the cue. This applies even when they are reasoning about social exchange, or social contract, situations that may or *may not* involve the possibility of cheating, depending upon whose perspective one takes in viewing the situation. To test this, researchers devised social contract situations and rules in which both parties can cheat (bilateral cheating situations) and social contract situations and rules in which only one party can cheat (unilateral cheating situations). Subjects who were cued to either party's perspective in the bilateral cheating situations were equally good at following the given rule. But in the unilateral cheating situation, only those subjects who were cued to the perspective of the party who could be cheated were good at following the rule. See, for example, Gerd Gigerenzer and Klaus Hug, "Domain-Specific Reasoning: Social Contracts, Cheating, and Perspective Change," *Cognition* 43 (1992): 127–71, esp. 153–64.

To illustrate this aspect of perspective change, consider the following rule:

> Rule 7: If an employee works on the weekend, then that person gets a day off during the week.

Notice that in this situation an employee can cheat his or her employer by taking a day off without having worked on the weekend, or an employer can cheat his or her employee by not giving the employee a day off after the employee has worked on the weekend. This is therefore a social contract rule that is open to bilateral cheating. Subjects were equally good at detecting violations of this rule regardless of whether they were cued to the perspective of an employer, looking for employees who cheated, or employees, looking for employers who cheated. But now consider this rule:

> Rule 8: If a passenger is allowed to enter the country, then he or she must have had an inoculation against cholera.

According to Gigerenzer and Hug, this is an example of a social contract rule that is only open to unilateral cheating. A passenger can cheat the "country" (or the immigration officer of that country) by entering the country without an inoculation. But the country (or the immigration officer of the country) cannot cheat an inoculated passenger by not letting him or her into the country, because the rule does *not* say that if one receives an inoculation one must be let into the country. An inoculation is a necessary but not sufficient condition for entry. Subjects who were cued to the perspective of an immigration officer looking at four cards ("inoculated," "not inoculated," "allowed to enter," "not allowed to enter") and trying to ferret out cheaters were significantly better at determining violators of the rule than subjects who were cued to the perspective of first-time travelers to the country curious about whether they needed an inoculation and trying to determine this from an inspection of the four cards. This research may show that we reason better about social contract situations than about other situations *only* when we perceive that the social contract situation in question is open to the possibility of cheating. That is what I have tried to stress in this chapter.

The second point I want to emphasize is that what constitutes a "cost" or a "benefit" can be understood in many different ways. Thus, while detecting cheating in social contract situations does involve a reckoning of costs and benefits, it need not be the case that the situation in question is one in which material goods are exchanged. Geoffrey Miller makes this point in relation to the above example about the tattoo and the cassava root. Miller correctly points out that this situation is complicated somewhat by the fact that the "cost" for being able to eat cassava root is *not* really having a tattoo. Rather, the cost is a *requirement*: being married, which is signified by getting a tattoo. (This was indeed a complicated, and I think, poorly written, example to use on test subjects.) As I said, the subjects given the version that cued them to the fact that individuals could cheat by eating something they were not allowed to eat did better than their counterparts who were given a different explanation. Miller argues that this result is possible only if the group that did better reasoned that being married was a "cost," *at least in this context*, even if being married had significant benefits (one of which was being allowed to eat cassava root). See Geoffrey Miller, *The Mating Mind: How Sexual Choice Shaped the Evolution of Human Nature* (New York: Anchor Books, 2000), 302–3. I think Miller's analysis is correct. But none of it vitiates my argument in this chapter about reciprocal altruism.

20. It turns out that there is some degree to which familiarity helps one detect violations of descriptive rules. Individuals could detect violations of *familiar* descriptive rules about half as well as they could detect violators of *unfamiliar* social exchange rules. See Cosmides and Tooby, "Cognitive Adaptations for Social Exchange," esp. 184–87.

21. See ibid., esp. 193–95.

22. See ibid., 166.

23. For Kant's explication and discussion of the categorical imperative, see Immanuel Kant, *The Critique of Practical Reason*, trans. Lewis Beck (Indianapolis: Bobbs-Merrill, 1956), and *Foundations of the Metaphysics of Morals and What Is Enlightenment?* trans. Lewis Beck (Indianapolis, Ind.: Bobbs-Merrill, 1959).

24. See, for example, Friedrich Nietzsche, *On the Genealogy of Morals and Ecce Homo*, in *Basic Writings of Nietzsche*, trans. Walter Kaufmann (New York: Modern Library, 2000), esp. 484–88.

25. A sign indicating that the lines in question were reserved for customers with ten items or less would, of course, be grammatically incorrect. Still, this is probably what most supermarket signs of this type say. The sign at my local supermarket, however, gets it correct: ten items or fewer.

26. See Trivers, "The Evolution of Reciprocal Altruism," 49.

27. Very often, instances of cheating are not subtle at all. Cheaters in express checkout lines are not subtle. Neither are what I call "traffic cheaters." How often have you been backed up in traffic, waiting patiently to merge into, say, an off-ramp

on your right that is itself backed up? On your left, traffic is speeding by, while you and the others in your lane wait patiently. But then you notice that a car that has just sped past you on the left is now trying to merge into your lane, very far ahead of you, parallel to the very entrance to the off-ramp that you and others in your lane have been waiting patiently to enter. You would like everyone ahead of you to close the gap between their cars, so that this cheater has no possibility of merging into your lane. But there is often a "soft-hearted" soul who will let this cheater in, even though that hurts everyone behind the cheater.

28. See David Zucchino, *The Myth of the Welfare Queen* (New York: Simon and Schuster, 1997), 65. Zucchino points out that Reagan and other conservatives tended to grossly exaggerate the amount of money that any given "welfare queen" was making by defrauding the system. But, from the perspective of a reciprocal altruist, the dollar amount of the fraud is not really the point. Again, it is the principle that matters most.

29. William Jefferson Clinton, "Acceptance Address at the Democratic National Convention," *New York Times*, 17 July 1992, A 14.

30. See Thomas Jefferson, *Notes on Virginia*, in *The Life and Writings of Thomas Jefferson*, ed. Adrienne Koch and William Peden (New York: Modern Library, 1998), 238.

31. As I said, the last episode of "Little House on the Prairie" is quite odd. Perhaps it is just coincidence, but in that episode there is a strange similarity between the actions of the good people of Walnut Grove and the actions of Howard Roark, the architect-hero of Ayn Rand's novel *The Fountainhead*. In the novel, Roark finds out that a housing project he has designed was not constructed to his specifications. Before the project is officially completed he decides to dynamite the entire structure, rather than let what he sees as an architectural imperfection stand. See Ayn Rand, *The Fountainhead* (Philadelphia: Blakiston, 1943).

Conclusion: Brave New World Revisited—Again

1. But not always. Francis Fukuyama, an intellectual best known for his book *The End of History and the Last Man* and currently a member of the President's Council on Bioethics, opens his 2002 book *Our Posthuman Future* with a thought-provoking discussion of the dystopias of Huxley and Orwell. See Francis Fukuyama, *Our Posthuman Future: Consequences of the Biotechnology Revolution* (New York: Farrar, Straus and Giroux, 2002).

2. See Aldous Huxley, *Brave New World Revisited* (New York: Perennial Classics, 2000).

3. Ibid., 20.

4. Actually, something like this scenario is the plot of many a dystopia, includ-

ing Walker Percy's last novel *The Thanatos Syndrome* (New York: Farrar, Straus and Giroux, 1987). Percy's somalike substance is called "heavy sodium," which is probably a bit more like Prozac than soma, although this distinction may be overly fine. At any rate, Percy has his villains overdose on heavy sodium. Thus they become undone by their own creation.

5. See U.S. Code, Title 20, sec. 1401 (2001).

6. For example, the Paul Erlichs of the world may be said to speak for the prophets of doom, while the Julian Simons of the world may be said to speak for the prophets of progress. In fact, for several decades Erlich, a self described ecologist, and Dr. Simon, an economist of the laissez-faire school, carried on a spirited public debate about the effects of economic growth, until the latter's death in 1998. The debate, of course, continues today. One side claims that there are natural limits to growth, and that when those limits are exceeded humankind will destroy itself in famine, war, or eco-catastrophe. The other side claims that the genius of human innovation is our greatest "natural resource," and that it will always provide us with a solution to our problems if we but give it rein to do so. One side presents itself as composed of hardheaded realists who are adult enough to accept the fact that you cannot have everything in life, while portraying its opponents as technogeeks in the grip of a dangerously pollyannaish view of science and nature. The other side presents itself as forward-looking optimists who have at least a modicum of confidence in humanity as a whole, while portraying its opponents as dour pessimists who secretly yearn for an Edenic past that never existed. The question of which side is correct is not merely academic, nor is it likely to be settled anytime soon. But I cannot resist mentioning the outcome of a wager between Erlich and Simon that took place in 1980. Simon had always felt that Erlich's pessimism regarding the depletion of natural resources was unjustified, given the ability of humans to innovate and thus to replace one resource with another *in a way that would actually increase overall efficiency.* So, Simon challenged Erlich to a wager. Erlich could select any five natural resources, and Simon bet that the prices per unit of those resources (an obvious measure of their relative scarcity) would be *less* in ten years than they were when the wager was made. Erlich selected as his five scarce natural resources copper, chrome, nickel, tin, and tungsten. He then waited ten years to find that he had lost the bet. For a brief discussion of the wager, see "Iconoclastic Economist Julian Simon Dies," *Washington Post,* 11 February 1998, B6.

7. See *State v. Oakley,* 245 Wis. 2d 447, 629 N.W. 2d 200 (2001), reh'g denied per curiam 248 Wis. 2d 654, 635 N.W. 2d 760 (2001).

8. See *Skinner v. Oklahoma,* 316 U.S. 535 (1942).

9. U.S. Const. Amend. XVII.

10. See Aristotle, *Rhetoric,* trans. John Henry Freese, Loeb Classical Library (Cambridge, Mass.: Harvard University Press, 1982), 15.

Afterword: Writing on *The Blank Slate*

1. See Steven Pinker, *The Blank Slate: The Modern Denial of Human Nature* (New York: Viking Penguin, 2002).

2. Ibid., 85.

3. Ibid., 73.

4. Ibid., xi.

5. Ibid., 6.

6. Ibid., 139, emphasis in original.

7. Ibid., 143, the emphasis on "among" and "between" is Pinker's; the rest is mine.

8. Ibid., 145, emphasis in original.

9. Ibid., 147.

10. Ibid., 332.

11. Ibid., 180.

12. Ibid., 192–93.

13. Ibid., 237–38, emphasis in original.

14. See Plato, *Gorgias*, 460c, in R. E. Allen, *The Dialogues of Plato*, vol. 1 (New Haven, Conn.: Yale University Press, 1984), 245. There is much confusion about this quotation, and on this point generally. Plato is not saying that the just man will never *do* injustice. The just man may do an injustice if, for example, he is too weak to resist the *physical* temptation to, say, commit adultery. ("The spirit is willing but the flesh is weak," as St. Peter said.) But, in the context of the *Gorgias*, Plato makes clear that the just man is the man who has *learned* justice. Such a man could not desire to do an injustice because, in the full knowledge of the good, the desire to do its opposite would be irrational. Plato also makes clear that an "irrational will," like an "undirected will," is simply a contradiction in terms. To will is to will some *thing*. Thus the will, if it exists at all, must always have an objective toward which it is directed. This means that "free will" cannot literally exist. But this is not to be feared, since humans have the capability, through philosophy (or, now, science), to discover the proper object of the will. We can know what would make the best life possible.

ACKNOWLEDGMENTS

Every intelligent person who writes on the nature-versus-nurture debate understands that all humans are, of course, the unique products of their genes and their environments. I like to think that I got the best of both nature *and* nurture. My father, a mathematician by training, and my mother, a sociologist, provided the natural endowments that enabled me to take advantage of the nurturing environment that they provided—an environment especially rich in diverse intellectual stimulation. Robust discussions at the family dinner table were perhaps the single most important element constituting that nurturing environment. It is no exaggeration to say that my older brother and I grew up on such discussions—first as eager spectators and then, when we were able, as enthusiastic participants. It is also no exaggeration to say that such discussions, around a family table, are the sine qua non of a healthy family life; as a society we lose them at our peril. The discussions around our table encompassed science, art, politics, religion, and the whole scope of the human condition. It happened, of course, that in the fullness of time, first my brother, then I, left the family table, and the nurturing environment it constituted, to make a way in the world. In the course of my journey, I have sat at the feet of learned professors at some of the world's best universities; I have kept the company of philosophers and poets; and I have led my own students in conversations about the timeless topics that reverberate throughout the seminar rooms of the academy. But I have never known discussions more exhilarating, more intellectually rigorous, or more philosophically profound than the ones I enjoyed at my parents' dinner table. I thank them dearly for giving me a place at the table and a place in their hearts.

I also thank my parents for assistance of a more immediate sort. Both of them read a draft of the manuscript and provided extraordinarily helpful comments, many of which saved me from serious mathematical, scientific, or reasoning errors.

The deep influence of my brother's thinking is also palpable throughout this book. Before his untimely death in 1999, he helped me clarify several of the ideas in the book, particularly those in chapters five, six, and ten. He also helped me understand some of the finer points of biology—one of his majors at the Massachusetts Institute of Technology—and appreciate the connection between the human mind and the computer world of artificial intelligence— his lifelong passion after graduation. I deeply miss his wisdom and his counsel.

For challenging many of my arguments over the years, and thus for helping to keep me intellectually honest, I thank David Grassmick. Since our days together as first-year students at the University of Virginia, David has been an intellectual colleague, a good friend, and a trusted advisor.

I would also like to acknowledge the wisdom and the knowledge I have gained from my professors, and later my colleagues, at all of the various universities at which I have studied and taught, including the University of Virginia, Northwestern University, Truman State University, Queens College, Polytechnic University, New York University, and the College of Staten Island.

My colleagues at Baruch College, especially those in the Weissman School of Arts and Sciences and the School of Public Affairs, have been remarkably generous in providing support and encouragement over the course of the past six years, and also in answering specific questions within their disciplinary fields of expertise. I thank them and the staff at the college for helping to make this book a reality.

My students continue to be a resource for me, and a source of inspiration. Indeed, many of the ideas in this book have been developed and refined as a result of some very perceptive comments and criticisms by students in various courses I have taught over the years. I consider it both a privilege and a responsibility to have the opportunity, as a professor, to work with some of the finest young minds in the world.

I would also like to thank my research assistants: Dr. Roland Marden helped locate sources and added a critical perspective to many of my arguments; Tanya Silas took on the enormous task of expertly indexing the entire book, and made it look easy. Their contributions were immensely valuable. Additionally, I thank Danielle Ackley-McPhail and Jenelle Brooks for their helpful comments. Susan Finn has an eagle eye and a very generous heart. I am extremely grateful to her for putting the former in the service of the latter, and thereby making this book much better than it would have been.

For making my manuscript into a real book, all of the individuals connected with the Johns Hopkins University Press deserve more thanks than I can possibly convey. My editors, Dr. Trevor Lipscombe and Dr. Vincent Burke, could not have been more helpful. Dr. Lipscombe encouraged me from the beginning, and helped me to clarify my writing by always demanding more more *brevity*, that is. Dr. Burke's energy and enthusiasm made very good things happen. I would also like to thank Susan Lantz, who responded to all of my queries with a polite expertise that gave me confidence that my manuscript was in good hands. Several in-house editors also provided very useful comments about versions of the manuscript. When it comes to copyediting, it is simply impossible to imagine anyone better than Marie Blanchard. Her precision and attention to detail made me think that it must have been a computer that had copyedited my work; but her felicity with language, as she reworked my sometimes tortured syntax, made me realize that it was an artist after all.

I don't want to conclude without thanking all of my friends and family, here in the United States, in Trinidad, and elsewhere. I hope that everyone realizes how much I appreciate all of the love and support I have been given. That goes for nonhumans as well. I thank Ren and Carrots for their love.

It is de rigueur to place one's spouse at the end of one's list of acknowledgments. Often love and support, and especially patience, are acknowledged. Sometimes an author also mentions assistance of a more administrative nature. I readily acknowledge all of this help. Indeed, this book surely would never have been written had I not been given three full months to complete large parts of it at our country house in the Hudson Valley. These months were made possible through the Herculean organizational efforts of my wife, who relieved me of the burden of thinking about *anything* other than my book. But that having been said, the simple truth is this book was created out of passionate—often very passionate—discussions I have had with my wife, Lauralee. She is my first and best critic; my intellectual partner; my fount of wonder; and my one true love. *That* is what I am most thankful for, and what I want to acknowledge most strongly.

INDEX

adaptation, 11, 125, 133; seasickness and, 140–41; sexual selection and, 172; spandrels and, 144, 147–48

Adapted Mind: Evolutionary Psychology and the Generation of Culture, The (Barkow; Cosmides; Tooby), 3, 7, 118–19, 122, 138, 153, 174, 182

ADHD. *See* Attention Deficit Hyperactivity Disorder

agapē, Wilson and, 72

Alas, Poor Darwin: Arguments against Evolutionary Psychology (Rose & Rose), 77, 92

algorithms, as mental operations, 120–21, 123, 131

Allen, R. E.: Plato's notion of justice and, 278n. 14; Prisoner's Dilemma and, 203

alpha male, 2, 187

altruism, 2, 12, 151, 196–223; genuine sacrifice and, 203; not kin selection, 197–98; relation to truth, rationality, and self-interest, 201; *Sociobiology* and, 71; survival of the fittest and, 203–4. *See also* reciprocal altruism

altruist, 203, 219

altruistic: behavior and cheater detection, 215; behavior and kin selection, 197–98; commands of Christianity, 217–18

American Association for the Advancement of Science, 65

amour propre, 244

ancestral environment, 5, 11, 63, 122, 124–25, 135–36, 139, 142, 148; evolution of human eye and, 105–6; face recognition module and, 125; mind attribution module and, 126; rationality in, 185; reciprocal altruism in, 208; seasickness and, 140–41; spatial reasoning module and, 128–29

"Apportionment of Human Diversity, The" (Lewontin), 59

Aristotle, 104, 143; definition of rhetoric, 235

Aronowitz, Stanley, Sokal affair and, 10

artificial intelligence, 129–31. *See also* Turing, Alan; Turing Test

artificial selection, 103

Asimov, Isaac, 12

Atlantic Monthly. See Herrnstein, Richard

Attention Deficit Hyperactivity Disorder (ADHD), 13, 228

Aune, James Arnt, 262n. 45

autism, 13, 221, 229; mental modules and, 126–27

Axelrod, Robert, 206

Badcock, Christopher, 102, 169

Barkow, Jerome H., 4, 118, 153